AKADEMIE DER WISSENSCHAFTEN UND DER LITERATUR

ABHANDLUNGEN DER

MATHEMATISCH-NATURWISSENSCHAFTLICHEN KLASSE

JAHRGANG 1980 · Nr. 2

The Conchiolin Matrices in Nacreous Layers of Ammonoids and Fossil Nautiloids: A Survey

Part 1: Shell wall and septa

by

CHARLES GRÉGOIRE

With 7 textfigures, 59 figures on 15 plates and 3 tables

AKADEMIE DER WISSENSCHAFTEN UND DER LITERATUR · MAINZ
FRANZ STEINER VERLAG GMBH · WIESBADEN

CIP-Kurztitelaufnahme der Deutschen Bibliothek

Grégoire, Charles:

The conchiolin matrices in nacreous layers of ammonoids and fossil nautiloids: a survey / von Charles Grégoire. — Mainz: Akademie der Wiss. u. d. Literatur; Wiesbaden: Steiner

Pt. 1. Shell wall and septa. — 1980.

    (Abhandlungen der Mathematisch-Naturwissenschaftlichen Klasse / Akademie der Wissenschaften und der Literatur; Jg. 1980, Nr. 2)
    ISBN 3-515-03382-3

NE: Akademie der Wissenschaften und der Literatur ⟨Mainz⟩ / Mathematisch-Naturwissenschaftliche Klasse: Abhandlungen der Mathematisch-Naturwissenschaftlichen . . .

Vorgelegt von Hrn. Erben in der Plenarsitzung am 23. Februar 1980, zum Druck genehmigt am selben Tage, ausgegeben am 15. Januar 1981

© 1980 by Akademie der Wissenschaften und der Literatur, Mainz
DRUCK: HANS MEISTER KG, KASSEL
Printed in Germany

CONTENTS

1. Introduction 5
2. Materials and Techniques 6
3. Observations 8
   - 3.1. Mineralogic composition of the nacreous layers:
     Aragonite. Calcite. Other Minerals 8
   - 3.2. Architecture of the nacreous layer in the shell wall 10
   - 3.3. Ultrastructure of conchiolin. Modern *Nautilus*. Types of conchiolin alteration in fossil nacreous layers.
     Distribution of the types of conchiolin alteration in the shells of the different groups and ages 11
4. Discussion 18
   - 4.1. Techniques of preparation and reliability of the ultrastructural appearance of conchiolin in fossil nacreous layers 18
   - 4.2. Changes in the mineral and conchiolin components in fossil nacreous layers .. 20
     - 4.2.1. Mineral components 21
     - 4.2.2. Conchiolin matrices 21
     - 4.2.3. The different types of alteration of the nautiloid pattern. Mechanism of their production 22
     - 4.2.4. Preservation of the nautiloid pattern in fossil nacreous layers 24
   - 4.3. Biochemical composition of conchiolin in modern nacreous layers 28
   - 4.4. Biochemical composition of the remnants of fossil conchiolin 28
     - 4.4.1. Protein fraction of the conchiolin complex 28
     - 4.4.2. Polysaccharidic fraction of the conchiolin complex 30
     - 4.4.3. Trabeculae and fibrils in the conchiolin complex 31
     - 4.4.4. Types of alteration of fossil conchiolin and structural appearance of synthetic polypeptides 31
   - 4.5. Conchiolin ultrastructure, taxonomy and evolution 33
   - 4.6. Did the pattern of conchiolin ultrastructure in nacreous layers of ammonites differ from the nautiloid pattern? 33
5. Summary and Conclusions 35
6. Notes 39
7. References 42

8. Acknowledgements 52
9. Explanation of Figures and Plates 54
10. Tables 81

1. INTRODUCTION

This paper is a review of previous, and recent (unpublished) work on the ultrastructure of the conchiolin remains in nacreous layers of about 500 shells of Paleozoic and Mesozoic ammonoids and Paleozoic-Tertiary nautiloids. The data have been collected from observations in the transmission (TEM) and scanning (SEM) electron microscopes and from biochemical analyses of these organic remains.

Former conclusions about the mechanism of the alterations in the ultrastructure, the relation of these alterations to shell mineralogy, to modifications in the original shell architecture, to taxonomic classification and to stratigraphy have been re-examined.

Part I is concerned with the conchiolin of the shell wall and of the septa. Part II will deal with the organic remains of the structures involved in the buoyancy mechanism of the shells, namely the siphuncle and the brown membrane that coats the surfaces of the chambers.

2. MATERIALS AND TECHNIQUES

The mural and septal nacreous layers were collected in different regions of the shells.

The mineralogic composition of the samples was determined by X-ray powder diffraction analysis.

The samples were decalcified by chelation (saturated and 0.5 M aqueous solutions of the disodium salt of ethylene diamine tetraacetic acid, EDTA, Titriplex III Merck, Darmstadt, at pH 4.0 and 7.5). Before decalcification, the shells impregnated with oil (Buckhorn asphalt) were soaked in pyridine for several days (Prof. F. G. Stehli, personal communication, 1959). Benzene, ether and chloroform were not used because these solvents damage the conchiolin structures.

Fragments of nacreous layers were also decalcified in solutions of chromium (III) sulphate (Sundström and Zelander, 1968; Sundström 1968; Mutvei, 1970; Iwata, 1975a; Grégoire, Goffinet and Voss-Foucart, in the press). Other fragments were fixed and decalcified by the CPC method (Williams and Jackson, 1956, modified by Crenshaw and Ristedt, 1975): fixation overnight in 4 per cent formaldehyde −0.5 per cent cetylpyridinium chloride, then immersion in the same solution made 0.05 to 0.5 M with respect to EDTA, pH 8.0.

The interlamellar conchiolin sheets, collapsed and agglutinated after dissolution of the mineral lamellae, remain fastened together by their intercrystalline bridges. In order to break these bridges and to obtain individual sheets, the suspensions of conchiolin matrices were gently shaken, teased with needles or treated for a few seconds by ultrasonic irradiation (Headland Ultrasonic Equipment). Care was taken not to disrupt the individual interlamellar membranes themselves (see Section 3). After rinsing, suspensions of the residues were deposited onto copper mesh screens coated with films of formvar, air dried, or dried by the Anderson's critical point drying method, shadowcast with platinum, and examined in a Siemens (Elmiskop I) electron microscope, using a voltage of 80 Kv, a double condenser, a 200 microns condenser aperture, a 30 microns objective aperture and a cold stage.

Residues of interlamellar conchiolin sheets were also observed in their original topography in the form of pseudoreplicas, prepared by the double-stage replication method of Bradley (1954): these residues, freed by cleavage of the nacreous layers at the level of the interlamellar spaces, or by a stepwise etching of the layers polished in tangential orientation, parallel to the interlamellar spaces, cling to the intermediary plastic replica. These residues of conchiolin sheets are then transferred without disruption to the final, metallic (carbon-platinum) replica.

The architecture of the nacreous layers was examined in a scanning electron microscope (SEM : Stereoscan: Cambridge Scientific Instruments Ltd) on surfaces of transverse fracture and of polished and etched fragments of the nacreous layers. Direct replicas (single-stage, self-shadowed replication method: Towe and Cifelli, 1967) of the same surfaces were used for examination in the TEM.

The biuret tests were performed on aqueous suspensions of conchiolin shreds. The intensity of the positive lilac-violet reaction of these shreds was appreciated under a conventional microscope on drops of suspensions spread out into thin films between glass and coverslip.

3. OBSERVATIONS

In table 1, the nautiloids and the ammonoids have been ranged according to the classification used in the Treatise on Invertebrate Paleontology (Part K, Mollusca 4, Nautiloidea; Part L, Mollusca 4, Ammonoidea).

The remains of conchiolin of the nacreous layers have been studied in samples of 160 specimens from seven suborders of nautiloids, of 120 specimens from four suborders of Paleozoic ammonoids and of 220 specimens from 18 superfamilies of Mesozoic ammonoids.

Observations on 50 specimens of this material have been previously reported (Grégoire, 1959ab, 1966a, 1968, 1972bc; Voss-Foucart & Grégoire, 1971).

More than 14 000 electron micrographs (TEM) have been recorded.

3.1. MINERALOGIC COMPOSITION OF THE NACREOUS LAYERS

Crystallographic and X-rays analyses (Cornish & Kendall, 1889; Bøggild, 1930, Mayer, 1932; Stehli, 1956; Reyment & Eckstrand, 1956; Turekian & Armstrong, 1961; Hallam & O'Hara, 1962; Grandjean, Grégoire & Lutts, 1964; Palframan, 1967; Yochelson, White & Gordon, 1967; Hall & Kennedy, 1967; Erben, Flajs & Siehl, 1969) have shown different grades of preservation of the original aragonite in nacreous layers of ammonoids and fossil nautiloids.

The results of X-ray analysis of 180 specimens listed in table 1 have been reported in a previous paper (Grandjean & al., 1964).

Aragonite

The iridescence of the nacreous layer in the shell of the modern *Nautilus* appears to be enhanced with strongly metallic hues in many Cretaceous nautiloids and ammonoids. The oldest material listed in table 1 having retained their original aragonite were the Silurian Nautiloid *Tragoceras falcatum* (491) (Grandjean & al., 1964), the 350 million-year old Lower Carboniferous goniatite *Beyrichoceratoides* (662): (Hallam & O'Hara, 1962; Grandjean & al., 1964) and several Carboniferous (Mississippian and Pennsylvanian) nautiloids and ammonoids, including shells buried in Buckhorn asphalt (Stehli, 1956; Grandjean & al., 1964). Aragonite was also found in the Triassic *Carnites floridus* (965), in the Jurassic *Lytoceras* sp. (489), *Ammonites lineatus penicillatus* (785), *Psiloceras planorbis* (567, 988), *Androgynoceras lataecosta* (837), several *Amaltheidae* (408, 501, 795, 962), *Leioceras opalinum* (362, 367), *Harpoceras mulgravium* (452), 6 Lower Oxfordian *Stephanocerataceae* and 7 *Perisphinctaceae*.

Aragonite was highly predominant in the Cretaceous and Tertiary (Eocene, Oligocene and Miocene) shells.

In some samples of the shells listed above, recording of weak or very weak principal ray (3.03Å) of calcite indicated that these samples, including some from the Buckhorn asphalt, contained secondary calcite.

Calcite

Secondary calcite characterized the mineralogy of most Ordovician, Devonian, Permian and Jurassic shells examined in this study. The shells of two Lower Cretaceous ammonoids (1120, 1121) and of six Eocene nautiloids (1119, 1122 to 1126), all from the highlands of Peru, were composed of secondary calcite (Fig. 3), and were a notable exception to the predominant aragonitic composition of the fossil nacreous layers from these ages.

Other minerals:
1. *Quartz* (SiO_2), in Paleozoic nautiloids and ammonoids, in the Russian Jurassic *Neumayria* (514) and *Garniericeras* (429).
2. *Gypsum* ($CaSO_4 - 2H_2O$) in a Senonian *Scaphites* from Greenland (364, 365).
3. *Carbonate Apatite or Dahlite* ($Ca Mg Na H 3)_3$, in the Ordovician *Dolorthoceras sociale* (677), in the Russian Jurassic *Neumayria* (514), *Virgatites* (511, 798), *Garniericeras* (429) and in the Lower Cretaceous *Protohoplites* (498). A spherulitic structure characterizes dahlite in mollusc shells (Bøggild, 1930; Pettijohn, 1957; Grégoire 1966a) (Figs 7 and 8).
4. *Navajoite* ($V_2O_5 - 3H_2O$), in the Ordovician *Dolorthoceras sociale* (357), in the Mississippian *Endolobus clorensis* (969) and in the Pennsylvanian *Pseudorthoceras knoxense* (530: cameral deposits).
5. *Iron sulfide* (FeS_2) in Devonian nautiloids and ammonoids from Büdesheim, Eifel, in the Lower Pennsylvanian *Pygmaeoceras* (1210), in the Jurassic *Cenoceras* (835), *Eoderoceras* (833) *Pseudamaltheus* (962), *Harpoceras* (524), *Quenstedtoceras* (466), *Cardioceras* (467), *Craspedites* (460), in the Cretaceous *Baculites* (502) and in several *Hoplitidae* (430, 459, 456, 479, 498).

6. BaSO4 in a Lower Jurassic *Eleganticeras* (832).

The mineral residues of decalcification of the relatively considerable amounts of material (about 5 to 20 grams) used for the biochemical analyses, contained occasionally substances (e. g. FeS2, SiO2, BaSO4) (Voss-Foucart and Grégoire, 1971) which had not been detected in the small amounts of substance (a few mgr. and less) used for X-ray powder diffraction analysis.

3.2. ARCHITECTURE OF THE NACREOUS LAYER IN THE SHELL WALL

As in the modern *Nautilus* (Appellöf, 1892–1893; Schmidt, 1923, 1924, 1928; Bøggild, 1930; Grégoire, 1962; 1972a c; Stenzel, 1964; Mutvei, 1964, Erben, Flajs & Siehl, 1969), the shell wall of the fossil nautiloids and of many ammonites consists of three layers (Hyatt, 1872; Böhmers, 1936; Flower, 1939; Hölder, 1952; Mutvei, 1967, 1970, 1972b; Grégoire, 1959–1979; Erben, Flajs & Siehl, 1969; Erben & Reid, 1971; Ristedt, 1971; Birkelund & Hansen, 1974; Bayer, 1975; Howarth, 1975): an outer prismatic layer (spheruliticprismatic layer in the modern *Nautilus* : Mutvei, 1972b), a median nacreous layer and an inner prismatic layer (Fig. 2) (formerly called "helle Schicht", a functional variety ("Abart") of mother-of pearl: Schmidt, 1923, Grégoire, 1962). Reports of absence of the outer or inner layer may be explained by exfoliations produced by weathering.

In the brickwork configuration of modern and well preserved fossil nacreous layers (Fig. 1), organic sheets (interlamellar membranes) alternate with mineral lamellae. The sheets are interconnected by bridges of intercrystalline membranes or cords which cross the lamellae between the tabular crystals of aragonite, arranged in a flagging (Schmidt, 1923; Grégoire, 1957, 1962; Wada, 1961; Hudson, 1968; Towe & Hamilton, 1968). The columnar stacking pattern (Vertikale Säuligeschichtung: Schmidt 1923), one of the three modes of disposition of the tabular crystals of aragonite in successive lamellae of mother-of-pearl, characterizes the architecture of mother-of-pearl in the modern *Nautilus* and in physically unaltered fossil nautiloids and ammonoids (Fig. 1.) (von Nathusius-Königsborn, 1877; Biedermann, 1902; Schmidt, 1923, 1924; Ahrberg, 1935; "crystal stacking": Grégoire, 1962, TEM; Birkelund, 1967, TEM; Erben, Flajs & Siehl, 1969, SEM; Mutvei, 1964, 1967, 1970, 1972[b], SEM; Wise, 1970, SEM; Ristedt, 1971, SEM; Erben, 1972, SEM).

Replacement of aragonite by calcite in shells of fossil molluscs occurs by two mechanisms: solution-recrystallization (Bathurst, 1964b) or direct inversion in situ by a solid-state reaction (Bathurst, 1964a b, 1971; Dodd, type 5, 1966). Inversion of aragonite in the nacreous layers is characterized by preservation of relics of the original aragonitic lamellation (Figs 4, 5 and 6), scattered among the coarse-grained mosaics of calcite crystals (Flower, personal communication, 1961; Hudson, 1962; Bathurst, 1964a b; Land, 1967; Grégoire, 1966a, 1972b). These mosaics form the secondary calcitic architecture of the fossil nacreous layers after inversion (Bøggild, 1930; Moore, Lalicker & Fischer, 1952; Hölder & Mosebach, 1950; Hölder, 1952; Bathurst, 1964a b, 1971; Dodd, 1966; Grégoire, 1966a, 1972b; Land, 1967; Erben & al., 1969). Coalescence of parts of the stacks into composite structures, in which piles of tabular crystals are fused into coarse blocks showing subhedral or euhedral facets, is one of the first steps of the structural modifications of mother-of-pearl during direct inversion (Grégoire, 1966a, 1972b: see disc: 4-2-1). (Figs 4, 5 and 6).

In the present material, direct inversion was the predominant or exclusive mechanism of transformation of aragonite: relics of the original lamellar configuration and of the columnar stacks of crystals were observed in specimens as old as Ordovician (*Isorthoceras sociale*, 677, *Faberoceras*, 697–699, Grégoire, 1966a).

3.3. ULTRASTRUCTURE OF CONCHIOLIN

Decalcification by EDTA of the nacreous layers of the shell wall in modern *Nautilus* and in fossil Cephalopods (ammonoids and nautiloids) leaves a soluble and an insoluble conchiolin fractions. The present and former observations were mainly made on the EDTA insoluble fraction.

Modern Nautilus
The EDTA insoluble conchiolin fraction of the nacreous layer consists of continuous, soft, highly iridescent and transparent sheets. These sheets (interlamellar membranes) appear in the TEM in the form of mosaics of polygonal fields, which are delimited by straight ridges or cords of intercrystalline conchiolin. These polygonal fields are the outlines of the tabular facets of the dissolved aragonite crystals between which the interlamellar membranes were originally sandwiched ("crystal imprints": Grégoire, 1959a b, 1962, 1966a; Grégoire & Teichert, 1965; "crystal scars": Mutvei, 1969). Within the polygonal fields, the interlamellar conchiolin membranes are composed of networks of sturdy, irregularly cylindrical or varicose, knobby cords or trabeculae, resembling rhizomes of the garden iris (Textfig. 1 and Figs 9 and 10). These trabeculae are studded with hemispheric protuberances. They delimit a broad, generally elongate fenestration ("nautiloid pattern": Grégoire, Duchâteau & Florkin, 1955; Grégoire, 1957, 1962; Mutvei, 1969; Iwata, 1975a). Intertrabecular membranes are displayed like bridges across the

Textfigs 1–8 are outline drawings of conchiolin trabeculae from the nacreous layers and of their debris in different nautiloids and ammonoids.

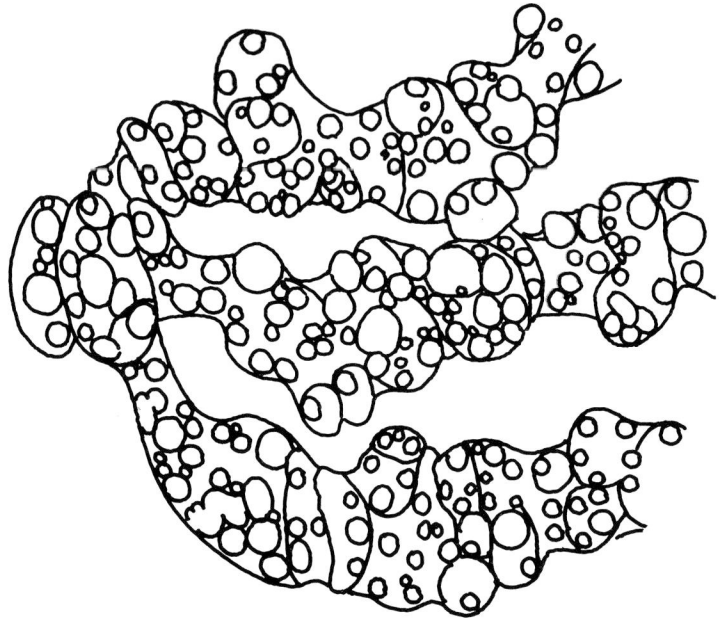

Textfig. 1: *Nautilus pompilius Linné.* (Modern). (209-14; 419-1). Tubercular shape of the sturdy, irregularly cylindrical, varicose trabeculae studded with bulging hemispheric protuberances of different sizes, and separated by an elongate fenestration (nautiloid pattern).

fenestration (EDTA material: Mutvei, 1969; Grégoire & Monty, in Grégoire, 1972c: Goffinet, Grégoire and Voss-Foucart, 1977; Chromium sulphate material: Iwata, 1975a b). As these membranes are extremely brittle, even in modern conchiolin, and frequently torn out during preparation of the sheets, they have disappeared in the fossil samples.

Types of conchiolin alteration in fossil nacreous layers

The conchiolin remnants of fossil nacreous layers are much scarcer in amount than in the modern *Nautilus*. They are brittle, soft or semi-rigid, yellow, grey, black or, more frequently, mahogany-brown shreds.

The biuret reaction of these conchiolin residues, observed under the conventional microscope (Table 1: asterisks), was consistently positive: the shreds appeared in the form of clusters of transparent polygonal flakes (Fig. 16), stained in dark-violet to pink-violet (lilac), with grades in intensity varying in the different samples. In a few samples, dark-violet or purple speckles marbled the uniformly lilac-coloured background of the flakes.

The differences in ultrastructure of the EDTA insoluble conchiolin fraction of the fossil nacreous layers have been classified into eight types. Transition forms were found between these types.

Electron diffraction diagrams of these types consisted of concentric rings with diffuse boundaries (Grégoire, 1966[a]).

Type 1: Networks of trabeculae ("reticulated sheets"), enclosed within polygonal fields delimited by intercrystalline cords and still presenting the features of the modern nautiloid pattern (Figs 13 and 17). In some samples the protuberances are no more visible and the trabeculae appear smooth.

In a few ammonites, the trabeculae are short, barrel-shaped, and delimit a polygonal rather than elongate fenestration (Figs 14, 36). As in the modern *Nautilus* (Grégoire, 1962), the pattern of conchiolin of the nacreous layers of the septa is more slender and the lattice-work denser than that of the mural conchiolin (Fig. 15).

Type 2: Bi – or tridimensional loose networks of irregularly cylindrical, slightly flattened, varicose, tubercle-shaped trabeculae, frequently broken into segments of different length, (Figs 19, 20, 21, 25, 26, 31, 32, 36, 37, 39, 40, 43) and found with alterations of type 5. In these trabeculae, swellings alternate with constrictions. Hemispheric tuberosities and spheroids assembled in clusters of botryoidal appearance (Textfigs 2a, c; 3b; 5a, b; 6a and Fig. 42) are bulging onto these trabeculae. The three-dimensional arrangement is produced by collapse, after decalcification, of the remnants of adjacent interlamellar sheets.

Type 3: Clustered or scattered spheroids or lenticular pebble-shaped particles, 20 to 200 millimicrons in diameter (Textfig. 2b, Figs 42, 44–49). In some preparations, these spheroids, which are fragments of trabeculae, have been preserved in their original location (Textfig. 2b, 7a) and the outlines of the networks are still recognizable.

Type 4: Larger, spheroidal, vesicle-like particles (from 200 millimicrons and up in diameter), composed of a relatively transparent, central spheroid, enveloped in an electron-denser bag-like membrane (Figs 25, 45, 51, 52). These particles, previously labelled papillae (Grégoire, 1959a) and knobs (Grégoire, 1966a, p. 6 and 15) reach considerable sizes (Fig. 25) (see also Grégoire, 1966a, Fig. 30).

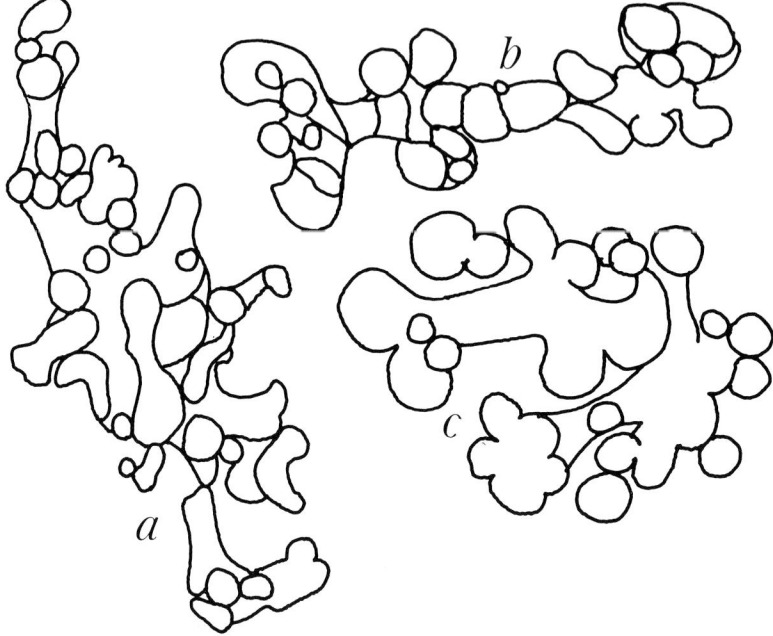

Textfig. 2: a. *Dolorthoceras sociale* (357 C) (Nautiloid) Upper Ordovician. b. *Clymenia* sp. (Ammonoid) (695) Upper Devonian. c. *Gastrioceras listeri* (Ammonoid, Goniatite) (667) Upper Carboniferous. Remnants of conchiolin trabeculae in the form of rounded, elongate, irregularly cylindrical segments (type 2), associated with spheroidal particles (type 3) forming clusters of botryoidal appearance. Age and conchiolin alterations seem to be unrelated.

Type 5: Bi – or tridimentional loose networks of flattened, irregularly broadened, twisted, angulate segments of trabeculae (Textfig. 4a and Figs 18, 24, 25, 27, 29, 30, 33), frequently associated with structures of Type 2 (Figs 20, 21, 25, 31, 32, 43).

Type 6: Fenestrate or continuous membranes (Textfig. 3a and Fig. 30) and their fragments in the form of discs, rounded or oval slabs (Fig. 34), flakes and rounded or indentated nodules (Fig. 38).

Type: 7: Feltworks of fibrils (Fig. 58).

Type 8: Elongate, soft laths or ribbons (Fig. 57).

The types of structure obtained after fixation and decalcification of the nacreous layers in Formaldehyde – cetyl – pyridinium chloride – EDTA

Textfig. 3: a and b: *Michelinoceras* sp. (1068–1, 1069) Upper Devonian. a: A considerable flattening of the trabeculae transformed the networks into perforated (f) membranes (type 6); b: In the same material, the alterations in the trabeculae appear in the form of cylindrical, twisted segments (type 2) and spheroidal particles (type 3). Identical alterations were found in fossils of different ages, e. g. in the Cretaceous *Eutrephoceras* (999) and the Miocene *Aturia australis* (1077).

(Fig. 43), did not differ from those of the EDTA insoluble conchiolin fraction.

In the organic remnants of nacreous layers decalcified by solutions of chromium sulphate, thick membranes were predominant over the other types of alteration of the trabeculae listed above (Fig. 35, and Grégoire, Goffinet and Voss-Foucart, in the press).

The different types of alteration were also recognized in samples dried by the critical point drying method of Anderson (Figs 22 and 23). However, compared with the samples dried in air, the structures of the samples dried by the critical/point drying method appeared to be considerably shrunk and more densely agglutinated (Fig. 22).

Distribution of the types of conchiolin alteration in the shells of the different groups and ages

The nautiloid pattern (type 1) is preserved in many conchiolin remnants of nacreous layers of Upper Pennsylvanian nautiloids (300 million-year-old) buried in asphaltic sandstones of Oklahoma (Buckhorn asphalt) (Grégoire,

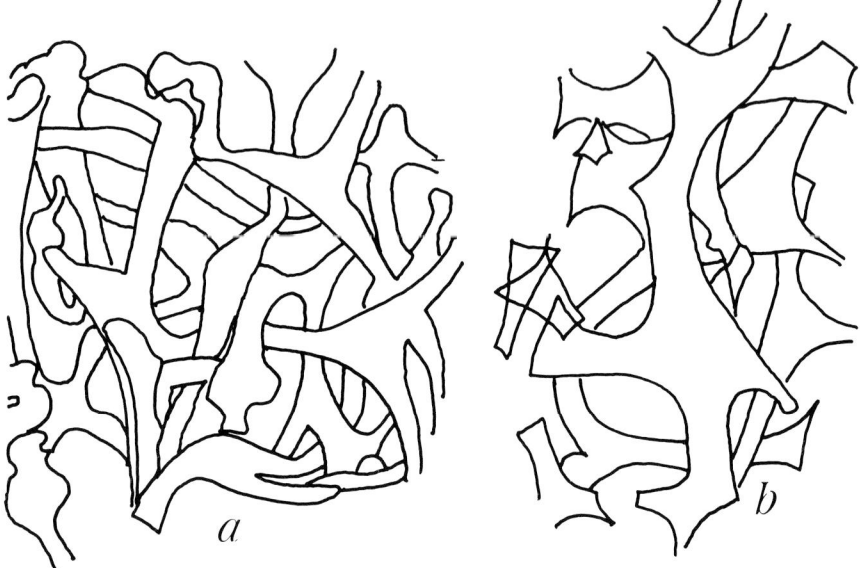

Textfig. 4: a: Loose networks of flattened angulate trabeculae were found in the Carboniferous *Hudsonoceras* and *Gastrioceras* (664, 667), together with other types of alteration (see Textfig. 3b); b: Identical alterations have been simulated by heating fragments of nacreous layer of the modern *Nautilus* in quartz tubes sealed under vacuum. (In this sample (811): dry, 400°C, for 30 minutes) (Grégoire, 1968).

1959b, 1966a; Grégoire and Teichert 1965; Ristedt, 1971). Fig. 1 (replica of a transverse fracture of nacreous layer) shows a system of interlamellar and intercrystalline membranes of conchiolin integrally preserved in an ammonoid of this horizon.

Features of the nautiloid pattern were also recognized in a few specimens from other stratigraphic ranges, e. g. in the Ordovician *Dolorthoceras sociale* (Grégoire, 1966a, Fig. 4), in a Jurassic *Amaltheus* (Fig. 41) in a Cretaceous *Scaphites* (Fig. 37), in the Paleocene *Aturoidea distans* (Fig. 26) and in an Oligocene *Aturia* (Grégoire, 1959a). The different types of alteration of conchiolin were scattered at random in the nacreous layers of Miocene to Ordovician specimens, still composed of their original aragonite, inverted to calcite, or recrystallized into other minerals. However, one or more types prevailed, but not consistently, in groups of shells or in different samples from single specimens. Loosened and disintegrated networks of cylindrical (type 2) and of flattened, angulate trabeculae (type 5) characterized hard, lustreless, honeybrown nacreous layers of Jurassic specimens composed of secondary calcite. Small spheroidal, pebble-shaped particles (type 3) were frequent in brittle, highly iridescent Cretaceous nacreous layers having retained their original aragonite, but altered into a porous substance. Similar particles (type 3) were predominant in the conchiolin matrices of the mainly aragonitic nacreous layers in a group of taxonomically different Pennsylvanian species (Brush Creek, Putnam Hill and Columbiana limestones, Allegheny and Conemaugh groups), some forming part of a thanatocoenose (Murphy, 1970) (Note 1).

In completely pyritized nacreous layers (e. g., in tables: orthocones and goniatites from Büdesheim, Eifel, 693, 695) and in those in which quartz had replaced aragonite (e. g. in tables: *Cravenoceras*, 1204 and *Anthracoceras*, 1205), the relatively scarce remnants of conchiolin consisted mainly of fragments of membranes in the form of thin, rounded or oval slabs (type 6) and of spheroids (type 3).

Fibers and fibrils (type 7) were found in several samples of nacreous layers contiguous to the inner surface of the shell wall and to the septal surfaces. Interpretation of the presence of these fibers will be discussed in section 4.4.2 and in Part II.

Elongate soft laths (type 8), originally located between the rod- and lath-shaped crystals radiating from the centre, were the organic remnants of the demineralized spherulites from the nacreous layers in which aragonite had been replaced by navajoite or carbonate apatite (dahlite) (e. g. in the Ordovician *Dolorthoceras sociale*, 357, 677) (Cloud & Barnes, 1957; Grandjean & al., 1964) (Note 2).

4. DISCUSSION

In the present and former studies on organic remains in fossil shells, the nacreous layers have been selected for the following reasons: 1. In several fossil nautiloids and especially in ammonoids, this layer represents an important or predominant part of the shell (Bøggild, 1930; Miller, Furnish & Schindewolf, 1957; Arkell, Kummel & Wright, 1957; Mutvei, 1967; Birkelund, 1967, Erben Flajs & Siehl, 1968).

2. Information on the ultrastructure of conchiolin of mother-of-pearl is considerable. The ultrastructure of the organic components of the prismatic layers in fossil cephalopods is not well known. In the present material, they seem to consist of sturdy, rigid membranes resembling palisades, with parallel folds, thickenings or rows of nodules. These parallel structures are possibly related to sheaths separating individual prisms. In fossil nautiloids and ammonoids, the prismatic layers are frequently more altered by weathering and contamination than the nacreous layers. They have disappeared in many specimens.

This and previous studies have shown that conchiolin residues of nacreous layers subsisted in all the samples from 500 ammonoids and fossil nautiloids in the range of Miocene (about 25 million-year) to Ordovician (about 450 million-year).

The problems of identification of these residues, of the mechanisms of their structural alterations, of the repercussions of the mineral environment (aragonite, calcite and other minerals) on these changes, discussed previously (Grégoire, 1966a), are reexamined below on the basis of a considerably extended material.

4.1. TECHNIQUES OF PREPARATION AND RELIABILITY OF THE ULTRASTRUCTURAL APPEARANCE OF CONCHIOLIN IN FOSSIL NACREOUS LAYERS

Modern conchiolin contains important amounts of EDTA- and water-soluble glycoproteins (Voss-Foucart, Laurent & Grégoire, 1969; Mutvei 1970;

Crenshaw 1972: about 15% in *Mercenaria*; Iwata, 1975b; Weiner & Hood, 1975).

Basic chromium (III) sulphate (Sundström & Zelander, 1968; Mutvei 1970) and Formaldehyde-cetyl-pyridinium-chloride-EDTA (CPC method of Williams & Jackson, 1956; Crenshaw & Ristedt, 1975) having fixing and decalcifying properties, are reported to prevent the loss of this water-soluble fraction. For this reason, we have tested these three methods of decalcification.

The ultrastructure of the conchiolin prepared by the CPC method did not differ from that of conchiolin freed by EDTA in modern (Goffinet, Grégoire & Voss-Foucart, 1977) and in fossil shells (Grégoire, Goffinet & Voss-Foucart, in the press). It seems then that elimination of the soluble fraction by EDTA did not distinctly modify the appearance of the trabeculae.

An interpretation of the miscellaneous structural appearances of conchiolin freed by chromium sulphate is difficult, because of the complexity of the still insufficiently known reaction between this decalcifier and the hard tissues associated with organic substances. Sundström and Zelander (1968) noted the possibility of a definite loss of an acid-soluble fraction, at least as regards human enamel decalcified by chromium sulphate. Substantial membranes are frequent in the structure of the modern (Iwata, 1975c) and fossil conchiolin (Fig. 35 and Grégoire & al., in the press) of nacreous layers decalcified by chromium sulphate. These membranes are associated with the other types of structure found in the EDTA-insoluble fraction. TEM observations on ultrathin sections of interlamellar membranes (Goffinet, Grégoire & Voss-Foucart, 1977) have shown that chromium sulphate produces a distinct shrinkage of the reticulated sheets of conchiolin, namely obliteration of the fenestration and agglutination of the trabeculae into membranes. Coarse membranes, out of proportion in size with that of the conchiolin membranes, might be precipitates produced by combination of chromium sulphate and calcium salts freed by the decalcifier. In any case, none of the structural appearances recorded in the material decalcified by chromium sulphate indicates a preservation of specific structural elements of conchiolin that might have been lost during decalcification by EDTA (Grégoire & al., in the press).

After desiccation by the critical point drying method (CPD), the fossil conchiolin structures did not differ from those observed after direct drying in air. However, the CPD method shrank all the structures, an artifact also reported in other organic materials (Boyde, Bailey, Jones & Tamarin, 1977; Billings-Gagliardi, Pockwinse & Schneider, 1978).

In conclusion, use of different procedures of preparation indicates that the EDTA insoluble conchiolin residues observed in this and previous papers are representative of the fossil conchiolin matrices. Some mechanical disruption of

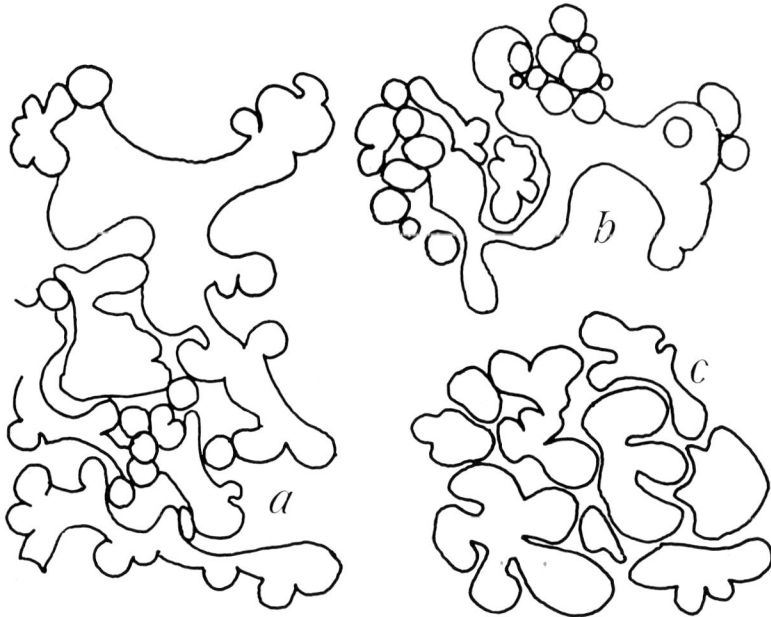

Textfig. 5: In two Jurassic (a: *Stepheoceras*, 989-2; b. *Hildoceras*, 1008) and c. a Cretaceous *Scaphites* (364-67), tortuous, cylindrical trabeculae (type 2), clusters of spheroidal particles (type 3) and nodules (type 6) were the alterations in the conchiolin trabeculae in a part of the samples.

the brittle networks of trabeculae might be produced in suspensions deposited on to grids. However, the arrangement of the trabeculae observed after this procedure did not differ from that found in pseudoreplicas, in which the mechanically undisturbed networks of trabeculae have been transferred on to the grids with the metallic replica (see Section 2).

4.2. CHANGES IN THE MINERAL AND CONCHIOLIN COMPONENTS IN FOSSIL NACREOUS LAYERS

It has long been reported (von Gümbel, 1884; Quenstedt, 1929; Newell, 1953; Macdonald, 1956; Wray & Daniels, 1957; Chave, 1964; Grandjean & al., 1964; Kitano & Hood, 1965; Hall & Kennedy, 1967; Kennedy & Hall, 1967; Kennedy, 1969, Kitano, Kanamori & Tokuyama, 1969; Allen, Martin & Taft, 1970; Jackson & Bischoff, 1971; Kamiya, 1975; Bayer, 1975; Hare & Hoering, 1977) that interlayered organic matter (conchiolin matrices) protects the original aragonite in fossil mollusc shells and that the biochemical composition of these matrices influences the conversion of aragonite.

On the other hand, sheltering between mineral layers prevents the destruction of conchiolin, especially in mother-of-pearl (Grégoire, 1959ab, 1966a; Lowenstam, 1963) and of its protein constituents (amino acids: Abelson 1954, 1957, 1959; Hare & Hoering, 1977; Samata 1979).

4.2.1. Mineral components

The data on the mineralogy of the shells listed in table 1 support previous records (Grandjean & al., 1964; Hall & Kennedy, 1967). Differences in the horizons, localities, enclosing lithology or degree of shell preservation are the causes of a few divergences (e. g. *Promicroceras, Leioceras opalinum*: Hall & Kennedy, 1967).

Attempts at thermal simulation on nacreous layers of the modern *Nautilus* of the changes in mineralogy having occurred during fossilization gave information about the steps of direct inversion of aragonite to calcite (Grégoire, Gisbourne & Hardy, 1969: Grégoire, 1972b, Figs 34–43). As in the fossil material, (see Section 3–2) the steps consisted of coalescence of contiguous, superimposed tabular crystals of aragonite in the stacks into coarse, polyhedral, anhedral or subhedral blocks, and final formation of mosaics of calcite crystals. As in the nacreous layers of the fossil shells, relics of the original columnar microarchitecture were scattered in the experimental material among the mosaics of calcite crystals.

4.2.2. Conchiolin matrices

From this and previous studies (Grégoire, 1959ab, 1966a, 1968, 1972abc; Grégoire & Lorent, 1971; Voss-Foucart & Grégoire, 1971), the following conclusions can be drawn about the modifications of structure of conchiolin in the nacreous layers of the fossil shells:

a. The modifications of the conchiolin matrices occurred relatively rapidly after burial, in still aragonitic nacreous layers.

b. The nature of the alterations was unrelated to the mineralogic composition of the nacreous layers (aragonite or calcite). Inversion of iridescent aragonite to harder, lustreless, honey-brown or white calcite did not seem to alter more deeply the already modified conchiolin structures. In some shells (e. g. the Jurassic nautiloid *Cenoceras*, 949, Table 1), in which still aragonitic regions are contiguous to regions transformed into secondary calcite, the appearance of the conchiolin matrices did not differ.

The observations did not support former statements about disappearance of the organic components in fossil shells older than Miocene, and in all the recrystallized shells from earlier ages. Inversion did not destroy the conchiolin (Hudson, 1962, 1967; Grégoire, 1966a). Conchiolin is even structurally less altered in Paleozoic and in Jurassic shells inverted to hard calcite than in some soft, porous nacreous layers of aragonitic, Cretaceous and Tertiary shells (Hudson, 1967; Grégoire, 1966a).

c. Absence of a distinct gradation in the conchiolin alterations in relation to age, from 25 million-year-old Miocene to 450–500 million-year-old Ordovician shells, indicates an early stabilization of the changes in conchiolin ultrastructure.

d. Experimental reproduction, by pyrolysis of modern nacreous layers, of the different ultrastructures found in fossils (Grégoire, 1964, 1966[b], 1968, 1972b; Grégoire & Lorent, 1971) (Figs 28, 50, 53, 54) suggested that elevation of temperature was the main factor of these alterations. As in fossils, the thermal changes in the conchiolin matrices developed rapidly, in still integrally or predominantly aragonitic nacreous layers, and were stabilized before inversion of aragonite to calcite.

4.2.3. The different types of alteration in the nautiloid pattern. Mechanism of their production

It has long been recognized, and recently on the basis of measurements of microhardness, tensile strength, compressive strength, bending strength, elasticity, fracture and rupture tests (Biedermann, 1914; Schmidt, 1928; Taylor & Layman, 1972; Currey & Taylor, 1974) that mother-of-pearl is the hardest (after the crossed lamellar structure, according to Taylor & Layman, 1972, Currey, 1976) and the strongest structure in mollusc shells. Hardness is likely to be correlated with resistance to abrasion and boring (Currey, 1976).

A pressure of 50 kilobars (about 50000 kgs/cm2), applied at room temperature for 5 hours on fragments of nacreous layers of the modern *Nautilus* shell (Grégoire, 1965, unpublished), produced only slight changes in the mineral architecture (note 3), and did not distinctly alter the pattern of conchiolin in the interlamellar membranes (Figs 11 and 12). External pressure (1.5 – 60 kilobars), associated with elevation of temperature (100° – 250° C) did not cause in the conchiolin matrices structural modifications that distinctly differed from those produced by elevated temperature alone (Grégoire & Lorent, 1971, note 4).

The great stability of mother-of-pearl also appeared in other laboratory experiments (Grégoire, 1968 pp. 6, 38, 39). Repeated boiling of nacreous layers in sea sediment (total boiling time: 53 hours) alternating with burial at 20°C for 42 months in the same environment (Grégoire, 1972b, p. 32) did not modify the architecture (Grégoire, 1972b, Figs 4 and 5) nor the nautiloid pattern of

Textfigs 6 and 7: (see discussion, section 4-4-4). Similarities between the structure of the fossil conchiolin residues and molecular aggregates of polypeptides are shown in outline drawings of Field Emission Sem micrographs recorded by Masuda, Sakurai and Osumi (1978) (about the same magnification).

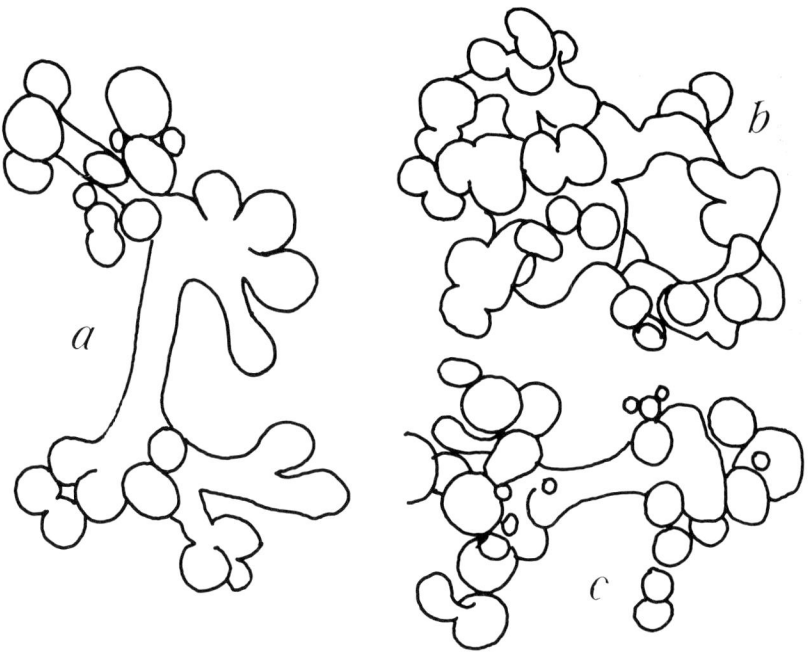

Textfig. 6: a: *Aturia* (323–335) sp. Eocene; b: Poly-L-Tyrosine (Plt) aggregates (spheroidal form) (Fig. 1a in Masuda & al., 1978); c: Poly-O-carbobenzoxy-L-Tyrosine (Pclt) aggregates (spheroidal form) (Fig. 1b in Masuda & al., 1978).

conchiolin. The trabeculae were only slightly shrunk (Grégoire, 1968, Figs 39–43).

Mother-of-pearl has larger amounts of organic matrix than other types of architecture (e. g. crossed lamellar, foliated, homogeneous structures: Grégoire, 1962; Taylor & Layman, 1972). According to Currey & Taylor, 1974, it is quite likely that this is important mechanically, and probably that organic matrix has an important role to play in determining the strength of mollusc shells.

On the other hand, as shown experimentally, (Grégoire, 1966b, 1968, 1972b; Voss-Foucart & Grégoire, 1971), conchiolin of nacreous layers is thermoresistant. It does not disappear structurally nor biochemically and it is still biuret-positive in samples of mother-of-pearl heated at 900° C for 5 hours.

Hardness and strength of its mineral components, thermoresistance of its organic components are exceptionally favourable conditions for a relative preservation of mother-of-pearl for long geological periods.

The different types of alteration in the ultrastructure of the fossil conchiolin matrices were caused by physical and chemical changes in the linear interlamellar cavities of the rigid nacreous layers. Enclosed in these spaces, the fossil conchiolin matrices escaped the effects of external compression, at least at the beginning of diagenesis. Except at a limited depth in the areas of fracture (note 5), the nacreous lamellae are impermeable and protect the interlamellar and intercrystalline conchiolin from the initial, mostly biochemical factors of diagenesis, such as hydrolysis by percolating, postdepositional water, bacterial activity, oxidation. These factors were generally active at temperatures and pressures similar to those at the surface of the earth (Prof. R. G. C. Bathurst, personal communication). However, this impermeability of its environment did not protect conchiolin from later elevation of temperature, which is the most significant geophysical factor involved in the subsequent metamorphic stage.

Under the influence of elevation of temperature, evaporation of the water contained in conchiolin or in the form of inclusions trapped within the individual crystals of aragonite (Römer, 1903; Hudson, 1962, 1967, 1968; Towe & Thompson, 1972), together with the release of volatile compounds of thermal decomposition of the conchiolin, caused a certain degree of pressure in the interlamellar spaces (note 6). A temporary additional compression within these spaces (note 7) was produced by a volume increase of the aragonite crystals surrounding the conchiolin matrices during their inversion to calcite (Hölder & Mosebach, 1950; Bathurst, 1964ab, 1970).

This elevation of pressure in the interlamellar spaces might have caused the flattening, broadening and coalescence of the trabeculae into fenestrate and, in a further stage, into continuous membranes (types 5 and 6).

These conditions of micro-environment (pressure in the interlamellar spaces) seem to be inconsistent with the presence in the conchiolin remnants of swollen, varicose trabeculae (type 2), of vesicle-like structures, some of considerable size (type 4) and of the same structures in heated modern nacreous layers (Figs 53 and 54). This divergence would be especially obvious if the interlamellar spaces were linear throughout in the fossil nacreous layers. Now, in the architecture of modern and of fossil mother-of-pearl, voids are found at different places of these layers.

1. The interlamellar spaces are broader at their junction with the intercrystalline spaces, where voids can be found.

2. It has long been reported (Schmidt, 1923) that gaps exist in the not quite mature mineral lamellae which underlie the last, still growing lamellae. In these still incomplete lamellae, voids subsist around the stacks of aragonite crystals until the still growing stacked crystals arrive in close contact with those of the contiguous stacks and constitute the final flagging of polygons. This flagging forms the surface of the lamellae in the mature mother-of-pearl (see an excellent diagram in Erben, 1972, Textfig. 1 region f).

3. Voids, larger than in modern nacreous layers, have been reported by Hudson (1968) between aragonite crystals of a Jurassic Mytilid. These voids are also visible between the aragonite crystals of a Jurassic *Dactylioceras* (343, Grégoire, 1959a, Fig. 3). A similar, diffuse, slight enlargement of the interlamellar and intercrystalline spaces has been observed, without dislocation of the architecture, in aragonitic relics of inverted, pyrolyzed, modern nacreous layers (Grégoire, 1972b, Figs 92, 93, 95).

4. Voids are found, during and after inversion of aragonite, in the loose mosaics of calcite blocks, especially between anhedral crystals, in fossil and in pyrolyzed modern shells (Grégoire, 1972b, Figs 58, 60, 83).

The void interstices in the nacreous layers made possible the conchiolin alterations in the form of swellings (types 2 and 4).

The swellings are probably related to the well known thermal expansion of the proteins (Pollard, 1964; Bathurst, 1964ab; Land, 1967). It is also probable that the varicosities and the vesicles were produced by local accumulations (bubbles: e. g. visible in trabeculae of Pronorites: See note 6) of volatile substances resulting from degradation of the proteins within the trabeculae, and were freed by dislocation of these trabeculae into scattered vesicles.

The origin of the spheroids or pebble-shaped particles (type 3), free (Plate 13, Textfig. 7b), still associated in networks (Textfig. 7a) or forming clusters of hemispheric elements with a botryoidal appearance, which protrude on to the swollen cylindrical trabeculae (type 2: Textfig. 2a, c; 3b; 5a, b; 6a), has been previously discussed (Grégoire, 1966a, p. 15, Figs 39 and 42). These spheroids were simulated in conchiolin of pyrolyzed nacre of the modern *Nautilus* (Fig. 50 and Grégoire, 1968). In some samples (Fig. 46), they have a remarkably uniform size. The perfect spherical or ovoid outlines of many of these particles suggest that they are not artifacts of fragmentation of the trabeculae during the preparation of the material (which would give blocks with irregular and indentated shapes), but that rounding up of the conchiolin fragments has occurred during fossilization. It has been formerly suggested (Grégoire, 1959a, 1966a) that these spheroids were protuberances like those that bulge on the trabeculae of nacre conchiolin of *Nautilus*, (Textfig. 1, Figs 11 & 12) and had been thermally inflated and detached from the

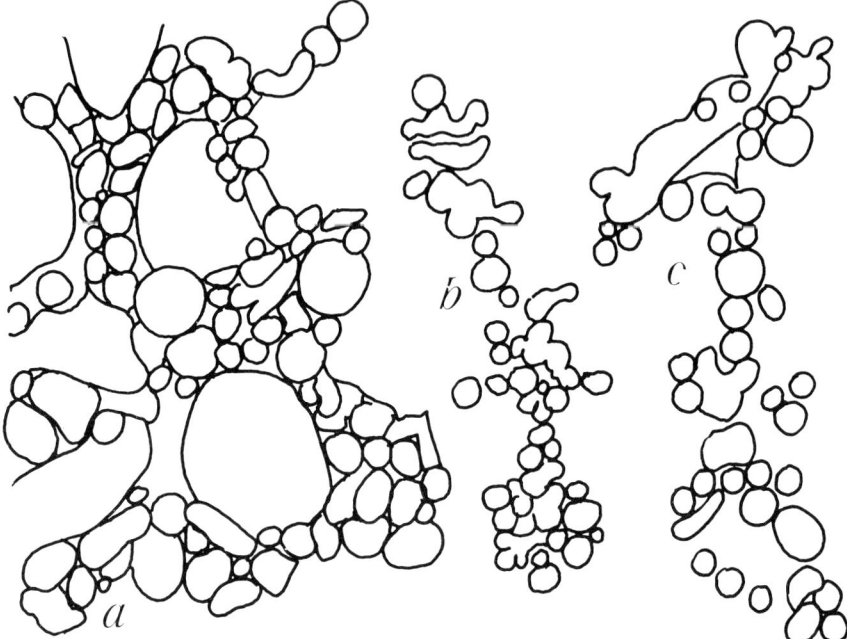

Textfig. 7: a: *Sphenodiscus* sp. (389), Cretaceous. Spheroidal and ovoid pebble-shaped particles (type 3) resulting from fragmentation without dislocation (see discussion, section 4-2-3) of the trabeculae. The outlines of the loosened networks are still recognizable. An identical alteration has been recorded for example in the Permian *Texoceras texanum* (Ammonoid, 858); b: Unidentified Nautiloid, Pennsylvanian (421–9–78) (septal material). Conchiolin residues predominantly in the form of chains of small, pebble-shaped, spheroidal particles (type 3); c: Identical structures are shown in micrographs of aggregates of poly-D L-Tryptophan (Pdlt). (Fig. 2a in Masuda & al., 1978).

trabeculae. The spheroids have been also considered as macromolecules of the conchiolin sheets (Wyckoff, 1972, p. 271).

Fibers and Fibrils (type 7) (see section 4.4.2.) are frequently associated with the conchiolin trabeculae in the modern (Grégoire & al., 1955; Goffinet, 1965, 1969; Mutvei, 1969, 1970) and pyrolyzed *Nautilus* shell (Grégoire, 1968). They were also recorded in the conchiolin remnants of ammonoid and fossil nautiloids (Grandjean & al. 1964; Grégoire, 1959, 1966a, 1967). However, in the fossil material listed in Table 1, fibrils were far less abundant than the other, ubiquitous types of conchiolin alteration. Transitions between altered trabeculae and fibrils were not seen. This scarcity in fiber- and fibril-like structures may be explained in two ways:

1. These fibrils, containing large amounts of chitin (Goffinet, 1965, 1969; Goffinet & Jeuniaux, 1969), have disappeared early during fossilization. This possibility will be examined below in connexion with biochemical data (see section 4.4.2.);

2. Fibrils, mixed with the other types of trabecular alterations, might be remnants of the brown membrane (Appellöf, 1892–1893), composed of dense feltworks of fibrils (Grégoire, 1962), and which, in the modern *Nautilus*, shell, coats the surfaces of the wall and of the septa in the chambers. During diagenesis, fragments of this membrane firmly adhered to these surfaces or were fused with the superficial layers of the inner part of the shell wall. They might have been incidentally included as "contaminants" in the samples of nacreous layers selected for decalcification. This possibility will be examined in Part II.

The electron-dense microfilaments contained in, or associated with, the remnants of conchiolin trabeculae in a few samples *(Cretaceous Baculites scotti,* 1003, *Hoploscaphites nicolleti,* 1005; Jurassic *Asteroceras* sp., 946, *Asteroceras stellare,* 929, Figs 39 and 40, *Echioceras* sp., 928, *Androgynoceras* sp., 828, *Harpoceras mulgravium,* 524, *Ludwigia murchisonae,* 836, *Cadoceras elatmae,* 796, *Garniericeras,* 402, *Oxynoticeras,* 942, *Ammonites lineatus,* 785) are possibly related to the micro-networks reported in ultrathin sections of trabeculae in the modern *Nautilus* conchiolin (Goffinet & al., 1977).

The Occurrence of biuret-positive remnants of conchiolin in completely or strongly pyritized shells, e. g. in Devonian orthocerids and goniatites from Büdesheim, Eifel, (Grandjean & al. 1964; Grégoire, 1966a; Clausen, 1968; Ristedt, 1971), (Tables, 693, 695 and Textfig. 2b) and in *Eoderoceras armatum,* 833, Lower Jurassic, indicates that parts of the interlamellar membranes escaped the conditions of production of pyrite: decomposition of conchiolin layers within the shell, giving rise to a local reducing micro-environment in the presence of sulfur and iron ions (Zobell, 1963; Taylor, 1964, Brown, 1966; Hudson, 1978).

The lath-shaped conchiolin structures (Type 8) freed by decalcification of the spherulites, of navajoite, or of dahlite (Fig. 58) were probably produced by moulding of the conchiolin remnants between the diverging crystalline rods of these spherulites (see note 8).

4.2.4. Preservation of the nautiloid pattern in fossil nacreous layers

Persistence of aragonite (Switzer and Boucot, 1955; Stehli, 1956; Hallam and O'Hara, 1962; Füchtbauer & Goldschmidt, 1964; Degens, 1967) and of the nautiloid pattern of conchiolin (Grégoire, 1959b, 1966a) in shells buried in the Lower Pennsylvanian Buckhorn asphalt has been explained by bituminous impregnation, shortly after entombment, of the Shells and of the surrounding sediments: This impregnation protected the shells from the early factors of diagenesis. However, the original proteins of this material have undergone a degradation comparable to that found in shells of other stratigraphic levels (Voss-Foucart & Grégoire, 1971, 1972: see section 4.4.). Physical preservation of organic matrix may therefore not coincide with its biochemical integrity (Grégoire, 1959a; Sacchi-Vialli, 1963; Hudson, 1967; Hare, 1969). In the Buckhorn asphalt material, this divergence suggests that the conchiolin proteins did not escape the action of elevation of temperature, whereas asphalt maintained the cohesion of the physical structures.

4.3. BIOCHEMICAL COMPOSITION OF CONCHIOLIN IN MODERN NACREOUS LAYERS

Conchiolin is a glycoprotidic complex.

Decalcification of mother-of-pearl leaves two organic fractions: a soluble fraction and insoluble fragments.

Successive extractions, from the insoluble fragments, of a water soluble fraction (soluble nacrin, frequently confused with the EDTA-soluble fraction) and of a scleroprotidic fraction (nacrosclérotine) leave a residual fraction (nacroin) (Grégoire & al., 1955, Goffinet, 1965, 1969, Voss-Foucart, 1970).

Further studies showed that nacroin alone (Goffinet, 1969) or combined with nacrosclerotine (Voss-Foucart, 1970) constitutes a complex composed of polypeptides (Grégoire & al., 1955) and polyaminosaccharides. Chitin represents 35 per cent of the nacroin fraction (Goffinet, 1969; Goffinet & Jeuniaux, 1969). Coarse fibers and thinner fibrils characterize in the TEM the structure of nacroin (Grégoire & al. 1955).

4.4. BIOCHEMICAL COMPOSITION OF THE REMNANTS OF FOSSIL CONCHIOLIN

The total amounts of organic constituents in fossil shells, especially amino acids, rapidly decrease with age, whereas the relative amounts of free amino acids (dialyzable soluble fraction) increase considerably (Abelson, 1954, 1959; Schoute-Vanneck, 1960; Florkin & al., 1961; Hare, 1962, 1969; Akiyama, 1964, 1971; Akiyama & Fujiwara, 1966; Ho, 1966; Mitterer, 1966; Hudson, 1967, 1968; Vallentyne, 1969; Akiyama & Wyckoff, 1970; Wyckoff, 1972; Bevelander & Nakahara, 1975; Weiner, Lowenstam & Hood, 1978).

4.4.1. Protein fraction of the conchiolin complex

The positive biuret reaction, recorded in the conchiolin shreds (Grandjean & al., 1964; Grégoire, 1966a) shows that the residues of fossil conchiolin are not mere carbonaceous substances, but physical substrates of preserved peptide bonds, still present in materials as old as the Lower Ordovician, 400–500 million-year-old specimens listed in table 1 (*Endoceras*, 793, 961, 454, *Armenoceras*, 701, *Orthonybyoceras*, 705, *Treptoceras*, 1050, *Leurocycloceras*, 698, *Dawsonoceras*, 792, *Dolorthoceras*, 357, 360, 677, *Faberoceras*, 696, 699, *Estonioceras*, 791, *Tragoceras*, 491: Grégoire, 1966a).

According to other data (Hare, 1969), the peptide bonds are hydrolyzed in geologically short times (10^4 to 10^6 years). In older fossils, the total amino acid concentration is essentially equivalent to the free amino acid fraction. Attempts at detection of any meaningful amino acids in the non-dialyzable fraction in other classes of fossils repeatedly failed. In all instances, the structural preservation was good (Towe, 1976, personal communication).

Analysis of the non-dialyzable conchiolin fraction after decalcification of the nacreous layers has been performed in 33 specimens of the material listed in table 1. (Florkin & al., 1961: 3 Oligocene-Eocene nautiloids; Voss-Foucart & Grégoire, 1971: 30 Miocene to Devonian nautiloids and ammonoids). In all the samples, hydrolysis of this fraction left assemblages of amino acids (Notes 9 and 10). On the basis of the reaction kinetics of amino acid compounds during fossilization and experimental diagenesis (Vallentyne, 1964; Drozdova, 1973), some amino acids of these assemblages might have been diagenetically formed by breakdown and transformation of other original amino acids.

In all the samples, the amino acid pattern was found to be strikingly uniform. The relative proportion of the amino acids differed from the biochemical pattern of conchiolin in the modern *Nautilus*. Like the structural modifications of conchiolin, this pattern did not significantly vary in the range of Miocene to Devonian, in still aragonitic nacreous layers or after their inversion to calcite. The pattern was found to be stabilized from the Miocene, the most recent age of the material available in these studies. A much earlier start of structural (Grégoire, 1959a, in *Holocene Unionidae*) and biochemical diagenetic denaturation of conchiolin has been reported in fossil Pelecypod shells (Hare, 1962; Mitterer, 1966; Tazaki, 1967; Kamiya, 1975; Bevelander & Nakahara, 1975; Cantaluppi & Rossi, 1976). Diagenetic changes are already recognizable after a few hundred years. There is an increased, step-wise degradation of the proteins between the Pleistocene and the Miocene (Ho, 1966), with a stabilization in the Middle and Upper Miocene (Hare & Mitterer, 1969; Matter & al., 1969), also observed in cephalopod shells (Voss-Foucart & Grégoire, 1971).

Previous work has shown that it is possible to simulate by heating modern shells, not only the structural alterations of conchiolin in fossils (see section 3), but also its biochemical diagenetic degradation (modern and Pleistocene

Pelecypods: Hare & Mitterer, 1969; Vallentyne, 1969; Hare, 1974; *Nautilus*: Grégoire & Voss-Foucart, 1970, Voss-Foucart & Grégoire, 1972, 1973ab).

Biochemical analyses of mother-of-pearl of *Nautilus* heated in the range of 150 °C to 900 °C showed a considerable reduction in the amounts of insoluble residues of conchiolin and persistence in these residues at all stages (including 900 °C), of a positive biuret reaction and of polypeptides with amino acid patterns similar to those recorded in Paleozoic, Mesozoic and Tertiary Cephalopods. As in fossils, the protein degradation was stabilized in the pyrolyzed material (Hare & Mitterer, 1969; Voss-Foucart & Grégoire, 1972, 1973a; Totten & al., 1973; Kamiya, 1975).

4.4.2. Polysaccharidic fraction of the conchiolin complex

Reports of persistence of chitin in fossils on the basis of reliable chemical methods of identification are scarce (Abderhalden & Heyns, 1933: in wings of Eocene Insects embedded in amber; Carlisle, 1964: in a *Cambrian Hyolithellus*). Chitin is considered as especially resistant to degradation by usual chemical agents.

However, chitin and its product of hydrolysis, glucosamine, were absent or in negligible amounts in the conchiolin remnants of the 33 Miocene to Devonian nacreous layers used for the biochemical analyses of the protidic fraction. The same negative results were observed in short-time pyrolysis of modern mother-of-pearl of *Nautilus* at high temperatures (300 °C to 900 °C) (Voss-Foucart & Grégoire, 1973a). On the other hand, heating of mother-of-pearl at lower temperature (160 °C) for long periods of time (15 days to nine months), without or under pressure (Voss-Foucart & Grégoire, 1973b, 1975), showed a selective preservation of the polysaccharidic fraction of conchiolin in the form of high amounts of glucosamine, firmly bound to substantial feltworks of fibrils. In this experiment, fibrils constituted the predominant structural component of the insoluble residue.

The absence or scarcity of glucosamine and of fibrils in fossil conchiolin of various ages and in short-term experiments of pyrolysis of modern nacre at high temperatures (Grégoire, 1966a, b, 1968; Voss-Foucart & Grégoire, 1971, 1973a) suggests that these materials are less resistant than certain peptide bonds and polypeptidic assemblages, which can be found even in Devonian (Grégoire, 1966ab; Voss-Foucart and Grégoire, 1971) and Ordovician (Grégoire, 1966ab) conchiolin remnants. In another interpretation about disappearance of chitin in fossils (Wyckoff, 1972, p. 117), polysaccharides with elementary particles smaller than those of cellulose have a tendency to break down into soluble sugars. This makes it improbable that they would be found in any but fairly recent specimens.

4.4.3. Trabeculae and fibrils in the conchiolin complex

The relationships between trabeculae and fibrils in conchiolin are still obscure. It has been previously suggested (Grégoire & al. 1955; Grégoire, 1960, 1967, 1972a; Goffinet, 1969; Goffinet & Jeuniaux, 1969), that the trabeculae of mother-of-pearl conchiolin might be composed of a fibrillar, polysaccharidic core, on to which polypeptides, removed by decalcifiers (leaching off), constitute a more or less labile muff.

Ultrathin, transverse sections of conchiolin trabeculae of the modern *Nautilus*, prepared by fixative-decalcifiers conservative of the fine ultrastructures (Goffinet & al., 1977) did not show such a fibrillar core. The basic ultrastructure of these trabeculae seems instead to consist of an extremely thin micro-network, uniformly dispersed throughout the trabeculae, except in spherical areas, perhaps related to the protuberances. The knots of this network are dense grains, 23–40 Å in average diameter. These knots are united by short, straight, thin organic connexions which delimit polygonal fields. It is not known whether these micro-networks belong to the protidic or to the polysaccharidic fraction of conchiolin.

Continuity between conchiolin trabeculae and fibers has been reported, namely at the junction of the septal neck with the connecting ring of the *Nautilus* siphon (Mutvei, 1972a, SEM; Grégoire, 1973, TEM). Transitions, by progressive thinning, between conchiolin trabeculae and fibrils, have been observed in modern mother-of-pearl heated at 160 °C for long periods of time (Voss-Foucart & Grégoire, 1975).

Resemblances between nacroin and fibrils have been reported in the conchiolin remnants of Eocene and Oligocene nautiloids (Grégoire, 1959a). Iwata (1975c) identified with nacroin the fibrils which constitute the main structural alteration of conchiolin in Pleistocene and Pliocene Bivalvia.

However, in Bivalvia, the "pelecypod pattern" of nacre conchiolin shows tight networks of slender, fibril-like trabeculae (Grégoire & al., 1955; Grégoire, 1960). Under these conditions, it is difficult to distinguish, on the basis of structural appearance only, intact from altered trabeculae, especially in disintegrated original networks.

4.4.4. Types of alteration of fossil conchiolin and structural appearance of synthetic polypeptides
(Textfigs 6 and 7; Fig. 56)

In the conchiolin trabeculae of the modern nacreous layers, the topography, at the molecular level, of the components of the protein-polysaccharidic complex, is still unknown (Grégoire, 1967; Goffinet & al., 1977). The magnification and power of resolution obtainable under the best conditions of the conventional microscopy are too low to make possible a precise location by the biuret reaction of the peptide bonds within the trabecular residues of the interlamellar membranes of conchiolin.

Recently, Masuda, Sakurai & Osumi (1976, 1978) have studied in the electron microscope the ultrastructure of precipitates of synthetic amino acids and polypeptide molecules. These assemblages, examined at about the same magnification, resemble the types of physical alteration of the remnants of fossil and modern, pyrolyzed, conchiolin, described in this and previous papers. For example (Textfigs 6a b c, Textfig. 7bc), a spherical form of molecular aggregates of polypeptides may be compared to the spheroids scattered, attached in chains or agglutinated into clusters of botryoidal appearance on to swollen cylindrical trabeculae (types 2 and 3).

Precipitates, in various solvents, of synthetic polypeptides, basically fibrous in character, appeared in the form of fibrils, entangled into ropes and bundles, or of larger structures such as convoluted strands, tortuous multistrands segmented into lath-like units, stout, open or closed hairpins, loops or toroids. Electron diffraction patterns were not obtained from these structures. Only the usual diffuse rings characteristic of amorphous matter were recorded (Blais & Geil, 1968).

Similar structures were found in the conchiolin residues of fossil (Grégoire, 1966a, 1968) and those of pyrolyzed, modern mother-of-pearl (Grégoire, 1968). Structures resembling closed loops or toroids are a form of pyrolytic conchiolin transformation in *Nautilus* mother-of-pearl heated for instance at 600 °C for 5 minutes in tubes sealed under vacuum (Fig. 56 and Grégoire, 1968).

The data reported above suggest that the types of alteration in the biuret-positive conchiolin residues of fossil nacreous layers might actually constitute the bulk of the physical substrate of the polypeptides recorded with biochemical methods. Under these conditions, the spheroids and pebble-shaped particles might represent biochemical compounds visible at the molecular level and not merely products of random disintegration.

Most of the recorded polypeptides are probably not the original assemblages of amino acids. As noted by Drozdova (1973), diagenetic partial breaking of original polypeptides into free amino acids might have been followed by interaction of the freed functional groups and synthesis of new polypeptidic compounds, a suggestion also made previously by Vallentyne (1964) (see section 4.4.1.).

4.5. CONCHIOLIN ULTRASTRUCTURE, TAXONOMY AND EVOLUTION

The nautiloid pattern of conchiolin in nacreous layers is highly representative of the modern genus *Nautilus*.

The features of this pattern have been preserved nearly unchanged in conchiolin of Pennsylvanian nautiloids (e. g. *Pseudorthoceras knoxense McChesney*: Grégoire, 1959b). They are recognizable in a few nautiloids from other horizons, including the Ordovician genus *Dolorthoceras sociale* (Grégoire, 1966a). The nautiloid pattern seems thus to have been stable for several hundred millions of years.

The material used in this and former studies includes specimens from genera which are important in evolution of cephalopods (*Armenoceras, Protocycloceras, Cyrthoceratites, Kionoceras, Cenoceras, Tornoceras, Manticoceras, Michelinoceras, Ophiceras, Psiloceras, Rhacophyllites*: see Kummel, 1956; Teichert, 1967). Some of these genera were the only survivors of former cephalopod faunas and ancestors of great lineages. In this material, no special structural feature of the conchiolin matrices could be recorded at the TEM level, which would distinguish these shells from many others without their importance in evolution.

Experimental simulation (Grégoire, 1964, 1966b, 1968, 1972b) in modern nacreous layers of the physical alterations of the conchiolin matrices in fossils (Types 2 to 8) showed that these alterations were produced by diagenesis, especially by elevation of temperature, and did not disclose extinct patterns nor represent taxonomic differences. Because of the early development of these types of alteration, of their random distribution in the same or different shells, of their stabilization throughout the further geological times, the ultrastructure of the conchiolin matrices seems to be useless in attempts at delineating taxonomic and evolutionary relationships in fossil shells (Voss-Foucart, 1970; Voss-Foucart & Grégoire, 1971).

Predominance of one type of alteration, reported for instance in the Pennsylvanian Putnam Hill assemblage of nautilids and ammonoids, forming part of a thanatocoenose, might be explained by burial in identical diagenetic conditions and not by original taxonomic differences.

4.6. DID THE PATTERN OF CONCHIOLIN ULTRASTRUCTURE IN NACREOUS LAYERS OF AMMONITES DIFFER FROM THE NAUTILOID PATTERN?

The Pennsylvanian shells from Buckhorn asphalt formation constitute an exceptionally favourable material for a comparison of the ultrastructure of the conchiolin matrices in ammonoids and in nautiloids, entombed throughout the geological times in identical environmental conditions.

This comparison was inconclusive. The greater part of the conchiolin matrices of the nacreous layers in the Buckhorn asphalt ammonites available in this and previous studies had undergone diagenetic changes. The modifications of their ultrastructure did not distinctly differ from the current types of physical alteration found in nautiloids and ammonoids from all ages and horizons.

However, in a few samples (Fig. 14; see also Grandjean & al., 1964, Fig. 2; Grégoire & Teichert, 1965, Figs 2 and 3, Pl. IX, Pl. X, Fig. 1; Grégoire, 1966a, Figs 30, 32 and 36), networks of short, dumpy, barrel-shaped trabeculae, delimiting a rounded, polygonal fenestration, might represent a pattern different from the nautiloid pattern. It might be objected that similar features are recognized in parts of the remnants of conchiolin networks of the giant nautiloid *Endolobus clorensis* (Figs 20 and 21).

Records of features of the nautiloid pattern in conchiolin of ammonoids from other horizons (e. g. in the Permian *Paraceltites elegans*, 856, Fig. 32, in the Jurassic *Amaltheus spinatus* Bruguière, 408, Fig. 41, and in the Cretaceous *Scaphites rosenkrantzi*, 364 – 367, Fig. 37) suggest that in some groups of ammonoids, the pattern of the conchiolin matrices might not much differ from the nautiloid pattern.

The slender septal nautiloid pattern was closely similar in ammonoids and nautiloids (compare Grégoire, 1966a, Fig. 31, unidentified ammonite, 754, with Fig. 15 of the present paper: *Pseudorthoceras knoxense*).

5. SUMMARY AND CONCLUSIONS

1. The data collected by electron microscopy and biochemistry on the remains of the conchiolin matrices in nacreous layers of ammonoids and fossil nautiloids have been reviewed.

2. The ultrastructure of the remnants of the conchiolin matrices has been studied in the transmission electron microscope in 160 specimens from seven suborders of nautiloids, 120 specimens from four suborders of Paleozoic ammonoids and 220 specimens from 18 superfamilies of Mesozoic ammonoids from stratigraphic horizons ranging from Miocene to Ordovician.

3. The nautiloid pattern, which characterizes the interlamellar conchiolin matrices of the nacreous layers of the modern *Nautilus*, has been recorded in a few specimens, especially in Pennsylvanian nautiloids, impregnated with asphalt (Buckhorn asphalt).

Persistence in Ordovician specimens of features of this nautiloid pattern of conchiolin and of relics of the original stacks of aragonite crystals suggests that the architecture of the nacreous layer of the modern *Nautilus* was already established 400–450 million-years ago and that more archaic patterns of shell ultrastructure in Nautiloidea must be searched for earlier in the geological scale.

4. In most fossil shells, the ultrastructure of the conchiolin matrices (networks of trabeculae) differed from the modern nautiloid pattern.

These differences, classified in several types, consisted of loosening of the networks, of flattening, swelling of the trabeculae, of their coalescence into membranes, and of their dislocation into spheroidal and vesicle-like particles.

5. Decalcifiers and fixative-decalcifiers, (EDTA, Chromium (III) sulphate, Formaldehyde – cetyl – pyridinium chloride – EDTA) freed identical conchiolin structures from fossil nacreous layers. Substantial membranes were predominant in the residues of decalcification by chromium sulphate of the conchiolin matrices. Drying by the critical point drying method of Anderson showed that the different types of ultrastructural alterations, are not artifacts of desiccation, but represent actual modifications of the trabeculae, developed during fossilization.

6. Simulation, on pyrolyzed modern nacreous layers, of the structural alterations observed in the residues of the fossil conchiolin matrices showed that these alterations were not related to original, extinct patterns of ultrastructure, but were diagenetic modifications, produced mainly by elevation of temperature (thermal factor of diagenesis), of a pattern identical or similar to that of the conchiolin matrices of the modern *Nautilus*.

7. None of these types of alteration characterizes a species or a taxonomic group. They are frequently found in the same sample or in different samples of a specimen.

8. Absence of gradation between Miocene and Ordovician in the structural changes of the residues of conchiolin indicates an early stabilization of these changes. An identical stabilization occurs in conchiolin from pyrolyzed mother-of-pearl of the modern *Nautilus* shell.

9. Inversion (by solid-state reaction) of aragonite to calcite, or replacement by other minerals, did not destroy physically the conchiolin. The ultrastructure of the conchiolin matrices did not distinctly differ in nacreous layers of specimens which had retained their original mineralogy (aragonite) and microarchitecture (stacking pattern of crystals) and in those inverted to calcite (mosaics of polyhedral crystals). Degradation of the conchiolin matrices and crystallographic composition of the nacreous layers are not distinctly related: the remnants of conchiolin were occasionally better preserved in a few calcitic, Paleozoic and Jurassic, than in still aragonitic, Cretaceous shells. In pyrolyzed modern nacre, as in the fossil shells, the alterations in conchiolin have already occurred when the nacreous layers were still integrally or predominantly aragonitic.

10. In regions of nacreous layers in which aragonite had been replaced by carbonate apatite (dahlite) or by navajoite, and the stack architecture by spherulites, the conchiolin remnants appeared in the form of sheaths, cords or laths, interposed between the elongate, radially diverging crystals, which compose these spherulites.

11. The biochemical literature differs about the grade of preservation of the original proteins in shells of fossil molluscs. Degradation of the original proteins into amino acids has been reported to occur in geologically short times (10^4 to 10^6 years). On the other hand, undialyzable assemblages of amino acids (polypeptides) have been found in the structural, biuret-positive remnants of conchiolin from nacreous layers of Miocene (25 million-year-old) to Lower Middle Devonian (about 390 million-year-old) shells. The biuret reaction (preservation of peptide bonds) was still positive in conchiolin from shells as old as Lower Ordovician (500 million-year-old).

Recent records of amino acid sequences and persistence of antigenic properties of the conchiolin remnants in Cretaceous shells indicate that a relative biochemical preservation of the original protidic components of conchiolin can be found in shells of much older ages than those previously reported.

Some types of structural alteration of the conchiolin remnants, shown in this and previous papers, resemble the ultrastructural aspects of synthetic polypeptides reported in a recent literature.

The undialyzable assemblages of amino acids recorded in fossil conchiolin are probably the products of degradation in a part of the original conchiolin proteins and of their subsequent reorganization during diagenesis into polypeptides different from the original biochemical structures.

12. Persistence of physical remnants of conchiolin in Ordovician shells and of assemblages of undialyzable amino acids in Devonian shells denote the remarkable thermoresistance of the conchiolin matrices in fossil nacreous layers. The same thermoresistance of conchiolin was observed in modern nacreous layers pyrolyzed at high temperatures.

13. Because of the early disintegration of the original conchiolin pattern and of the early stabilization of its alterations in fossil shells, it is doubtful that the ultrastructures of the conchiolin matrices in fossil Cephalopods could be used, in the present state of the methods, as additional criteria for resolving taxonomic and evolution problems.

6. NOTES

Note 1 This group was composed of the following species:
Nautiloids: A sample of an Upper Paleozoic orthoceracone related to *Bactrites* (1130: aragonite); *Kionoceras ungeri* (1136: aragonite); *Mooreoceras normale* (1133: aragonite); *Michelinoceras directum* (1139: aragonite). Nautilida: *Solenochilus* sp. (1131: aragonite); *Temnocheilus bellicosus* (1132: aragonite); *Domatoceras shepherdi* (1135: aragonite); *Liroceras liratum* (1137: aragonite); *Metacoceras macchesneyi* (1138: calcite).
Ammonoid: *Eothalassoceras* n. sp. (1134: aragonite).

Note 2 Similar types of conchiolin alteration have been also found in fossil shells from other orders: e. g. in Jurassic *Trigoniaceae*, in Cretaceous and Jurassic *Inoceramidae*, Pleistocene and Pliocene *Nuculidae* and *Mytilidae* (Grandjean & al., 1964), in Miocene, Oligocene, Cretaceous and Carboniferous Pelecypods and *Pleurotomariaceae* (Iwata, 1975c).

Note 3 Fragments of nacreous layers of modern *Nautilus*, compressed at 50 000 kgs per square cm. for 5 hours at 20 °C, have lost their faintly yellow-bluish iridescence, but not their brightness. They showed hues resembling those of 'turbot flesh'. In normal nacreous layers, the surfaces of the cross fractures, obtained by hammering, appear indentated and composed of iridescent, horizontally superimposed terraces, which are the regions of cleavage along the interlamellar spaces.

Fracture of the compressed fragments break them into blocks with smooth oblique slopes, inclined at angles of 60–70° to the horizontal surface of the nacreous layer. Direct replicas of the surfaces of cross fracture of the compressed fragments show persistence of the gross stack architecture, with possible coalescence of certain aragonite crystals in the lamellae. The tabular surfaces of the crystals of aragonite show parallel rows of tubercle-shaped elevations, orientated in two directions at an angle of 60–70° (see Fig. 59). The significance of these structural aspects of the compressed crystals of aragonite cannot be discussed within the scope of this survey.

In the nacreous layers, the crystals of aragonite have been considered (especially in Bivalves), as compact, single crystals, in mature mother-of-pearl

(Towe & Hamilton, 1968; Towe & Thompson, 1972), as complex crystals (Watabe, 1965, Erben, 1972, 1974), mosaic crystals (Wada, 1961), aggregates of lath-like subunits (Mutvei, 1970, 1972c), single crystals or polysynthetic twins (Erben, 1972), double pairs of crystals (Mutvei, 1977). The problem has been reviewed recently (Watabe & Dunkelberger, 1979, Carriker, 1978).

Note 4 The mineral components of pyrolyzed nacreous layers (e. g. between 300 °C and 400 °C) were brittle and cleaved spontaneously into mineral sheets of different thickness (Grégoire, 1972[b]).

The nacreous layers, heated under pressure (500 bars to 60 kilobars), were hard and remainded compact. In some experiments, inversion of aragonite to calcite occurred at higher temperature (600 °C – 800 °C) than in the merely heated controls (Grégoire & Lorent, 1971).

Note 5 Former experiments (Voss-Foucart & Grégoire, 1973b) suggest that the reactions of conchiolin differ in the central and peripheral parts of the fragments of *Nautilus* nacreous layers. Long-term pyrolysis (15 days to 9 months), at relatively low temperature (160 °C), of samples composed of a great number of fragments of modern mother-of-pearl immersed in water, left an EDTA insoluble fraction of conchiolin consisting of two kinds of structures: 1. networks of altered trabeculae, fused into membranes and 2. feltworks of fibrils. The former structures, probably from the central parts of the samples, without contact with the water filling the container, showed the thermal alterations of conchiolin formerly described (Types 2 to 5). On the other hand, the water, in contact for months with the portions of interlamellar membranes of conchiolin exposed on or near the peripheral surfaces of the fragments, might have extracted polypeptides from the conchiolin membranes, leaving only insoluble polysaccharidic residues in the form of fibrils (Fig. 57). A similar distinction between outermost and innermost samples has been made in water-leaching experiments on fossil bone specimens (Hare, 1974).

Note 6 Intense vesiculation within the trabecular cords of the pyrolyzed samples (Plate 14; see also Grégoire, 1968, Figs 28, 31, 84) and the release of gas bubbles by the conchiolin included in different layers of heated shells (e.g. *Pronorites*) support this interpretation.

Spontaneous explosion, accompanied by emission of pungent smell of tar, H2 S and NH 3, of quartz tubes sealed under vacuum and containing dry fragments of mother-of-pearl of very small size, heated at different temperatures (between 300 °C and 800 °C) (Grégoire, 1968, 1972b) suggests that the pressure developed by volatile substances released from the samples in the tubes, is not negligible (about 446 bars).

Note 7 This explanation concerns the conditions in the architectural stacking system of the aragonite crystals in mother-of-pearl of nautiloids. The conditions may differ in the brickwork architectural system (Backsteinbau) of Pelecypod nacreous layers (Schmidt, 1923; Grégoire, 1957, Fig. 12; 1972a, Fig. 9; 1972c Figs 1 and 2, TEM; Wada, 1961, Figs 33, 38, TEM; 1972, Fig. 12, TEM; Towe & Hamilton, 1968, Fig. 1, TEM.

Note 8 Squeezing of the conchiolin matrices by swollen crystals and subsequent remoulding of these matrices is illustrated in Fig. 55: many conchiolin residues of nacreous layers heated in sealed tubes at high temperatures (600 °C–900 °C) appear in the form of casts of euhedral facets of the polyhedral crystals of secondary calcite, between which the conchiolin remnants were compressed (see Grégoire, 1968, Figs 86, 92, 95 and 1972b, Figs 124 and 125.

Note 9 Measurement of the amino acid racemization in the protidic components of fossil shells makes possible to distinguish original proteins from modern protidic contaminants (see review in Bada & Schroeder, 1975). The racemization increases with the age of the fossil molluscs (Hare & Abelson, 1967). It does not occur in the shells of these molluscs in the absence of water (Hare & Mitterer, 1969; Hare, 1974; Hare & Hoering, 1977).

Weiner, Lowenstam, Hood (1976) have recorded original glycoproteins in the organic matrices of a late Cretaceous *Trigoniidae*, and Weiner, Lowenstam, Taborek & Hood (1979) have obtained important subfractions of shell matrix from a late Cretaceous *Baculites*. The authors explained these results by unusual conditions of preservation, such as conservation of the original shell microarchitecture, mineralogy and by absence of water. In these materials, racemization was extremely low or absent.

In the material selected for our analyses (Voss-Foucart and Grégoire, 1971: see also section 2-1), the racemization has not been measured. However, in a part of the samples, the conditions of preservation were identical to those reported by Weiner & al. In other samples, inversion of aragonite to calcite had occurred by direct solid-solid reaction, during which the concentration of indigenous water is reported to be very low. Under these conditions, the rate of racemization must have been very low.

The data on persistence of original polypeptides and on polycondensation of peptides and polypeptide remains during diagenesis of the fossil shells (Drozdova, 1973) are rapidly increasing (see a review of former literature in Voss-Foucart & Grégoire, 1971, in Wyckoff, 1972, and in Drozdova, 1973).

Experimental simulation, on dry nacreous layers of the modern *Nautilus*, of the biochemical changes recorded in fossils, showed persistence of polypeptides at temperatures as high as 900 °C (Voss-Foucart & Grégoire, 1972, 1973a, b).

Samata (1979) found still functionable Ca-binding peptides in Jurassic, Cretaceous and Miocene oyster shells. He noted (p. 85) that the presence of these peptides strongly suggests that they belong to the shells, because they have not been found in the protidic contaminants introduced into the shells during fossilization.

Detection of shell matrix components with preserved antigenic determinants (de Jong, Westbroek, Westbroek & Bruning, 1974; Westbroek, van der Meide, van der Wey-Kloppers, van der Sluys, de Leeuw & de Jong, 1979) also indicates that peptide bonds can still be present after geological times much longer than those at which they have been considered as completely hydrolyzed (e. g. 10^6 years) (Abelson, 1959: Hare, 1969).

Note 10 The presence of amino acids in the dialyzed residues of demineralization of the sedimentary rock surrounding the shell wall in Jurassic and Devonian samples (Voss-Foucart & Grégoire, 1971) seems to be inconsistent with the statement (see section 4-2) that mother-of-pearl is impermeable. However, the biuret-positive fragments of typical interlamellar sheets of conchiolin from which these amino acids had been obtained (see Fig. 17) were not originally free, but belonged to splinters of fracture exfoliated from the shell into the sediment. Except at their periphery, in the zones of fracture, from which amino acids could diffuse into the sediment, these splinters of fracture were as impermeable to water as the shell wall itself. It is evident that original polypeptides directly exposed to the percolating water during fossiliziation could not have persisted in a free state in the sediment (see disc. in Voss-Foucart & Grégoire, 1971).

7. REFERENCES

Abelson, P. H.: Organic constituents of fossils. Carnegie Inst. Washington, Yearbook n° 53, 97, Washington 1954.
— Organic constituents of fossils. In: Treatise on Marine Ecology and Palaeoecology. II. Palaeoecology. Geol. Soc. America, Mem. 67, 87–92, 1 Fig., 4 Tables, 1 Plate, 1957.
— Geochemistry of organic substances. In: Researches in Geochemistry, Vol. 1, P. H. Abelson, Ed., John Wiley and Sons, Inc. New York, 79–103, New York 1959.
Abderhalden, E. and Heyns, K.: Nachweis vom Chitin in Flügelresten von Coleopteren des oberen Mitteleocän. Biochem. Zeitschr., 259, 320, 1933.
Ahrberg, P.: Über den feineren Bau der Perlmutter von Schnecken und Cephalopoden. Arch. Molluskenk. 67, 1–20, 2 Plates, 1935.
Akiyama, M.: Quantitative analysis of the amino acids included in Japanese fossil scallop shells. Bull. Geol. Soc. Jap., 70 (828) 508–516, 4 Figs, 1 Table, 1964.
— The amino acid composition of fossil scallop shell proteins and non-proteins. Biomineralisation, 3, 65–70, 7 Tables, 1971.
— and Fujiwara, T.: Amino acid Survival of the Fossil Shells yielded from the Johmon Shellmounds in the Kanto District, Japan. Miscellan. Reports of the Res. Inst. for Natur. Resources, 67, 67–72, 1 Table, 1 Fig. 1966.
— and Wyckoff, R.W.G.: Total amino acid content of fossil Pecten shells. Proc. Natl. Acad. Sciences U.S.A., 67, 1097–1100, Washington 1970.
Allen, R. J., Martin, D. F. and Taft, W. H.: A study of the effect of selected amino acids on the recrystallization of aragonite to calcite. J. Inorg. Nucl. Chem. 32, 2963, 1970.
Appellöf, A.: Die Schalen von Sepia, Spirula und Nautilus-Studien über den Bau und das Wachstum. Kongl. Svenska Vetensk. Akad. Handlingar, 25 (7), 1–106, 3 Textfigs, 7 Plates, Stockholm 1893.
Arkell, W. J., Kummel, B. and Wright, C. W.: Mesozoic Ammonoidea. In: Treatise on Invertebrate Paleontology, ed. R. C. Moore, Part L, Mollusca 4. Cephalopoda, Ammonoidea, L 80–L 465, Figs 124–558, Geol. Soc. of America and Univ. of Kansas Press, Lawrence, 1957.
Bada, J. L. and Schroeder, R. A.: Amino Acid Racemization Reactions and their Geochemical Implications. Naturwiss., 62, 71–79, 3 Figs, 3 Tables, 1975.
Baccelle, L. S. and Garavello, A. L.: Prima segnalazione di Ammoniti Aptiane e Albiane nelle Dolomiti. Ann. Univ. Ferrara, N. Ser., IX, Sc. Geol. e Pal., IV, N° 7, 91–99, 1 Plate, Ferrara, 1967 a.
— — Ammoniti dei livelli cretacici di la Stua (Cortina d'Ampezzo). Ann. Univ. Ferrara, N. Ser., IX, Sc, Geol. e Pal., IV, N° 9, 117–153, 3 Plates, Ferrara 1967b.
Bathurst, R. G. C.: Diagenesis and Paleoecology: A. Survey. In: Approaches to Paleoecology (J. Imbrie and N. D. Newell, eds.), (John Wiley & Sons, Inc.), 319–344, New York 1964a.

- The replacement of aragonite by calcite in the molluscan shell wall. In: Approaches to Paleoecology (J. Imbrie and N. D. Newell, eds), (John Wiley & Sons, Inc.), 357–376, 8 Plates, New York 1964b.
- Problems of Lithification in Carbonate Muds. Proc. Geol. Assoc., 81 (3), 429–440, 1970.
- Carbonate Sediments and their Diagenesis. Developments in Sedimentology 12, Elsevier Publishing Company, Amsterdam, London, New York, 620 pages, 359 Figs, 1971.

Bayer, U.: Organische Tapeten im Ammoniten-Phragmokon und ihr Einfluß auf die Fossilisation. N. Jhrb. Geol. Paläontol. Monatsh., H 1, 12–25, 4 Figs, Jahrg. 1975.

Bevelander, G. and Nakahara, H.: Structure and amino acid composition of pearls exposed to sea water for four hundred years. Earth Sc. (Japan), 29, 87–91, 2 Tables, 3 Plates, 1975.

Biedermann, W.: Untersuchungen über Bau und Entstehung der Molluskenschalen. Jena. Zeitschr. Naturwiss. 36, N.F. 29, 1–164, 1902.
- Physiologie der Stütz- und Skelettsubstanzen. In: H. Winterstein, Handbuch der vergleichenden Physiologie, Bd. III, Teil I, 319–1188, Jena 1914.

Billings-Gagliardi, S., Pockwinse, Sh. M., and Schneider, G. B.: Morphological changes in isolated lymphocytes during preparation for SEM: Freeze Drying versus Critical-point Drying. Amer. J. Anat., 152, 383–390, 1978.

Birkelund, T.: Ammonites from the Upper Cretaceous of West Greenland. Meddel. om Grønland, 179, n° 7, 1–192, 125 Textfigs, 49 Plates, 1965.
- Submicroscopic shell structures in early growth-stages of Maastrichtian ammonites (*Saghalinites* and *Scaphites*). Meddel. fra Dansk. Geol. Foren, 17 (1) 95–101, 3 Figs, 4 Plates, København, 1967.
- and Hansen, H. J.: Shell Ultrastructures in some Maastrichtian Ammonoidea and Coleoidea and their taxonomic implications. Kongl. Danske Videnskab. Selsk. Biol. Skrifter, 20, 1–34, 7 Textfigs., 16 Plates, 1974.

Blais, J. J. B. P. and Geil, Ph.: Fibrillar polypeptide Aggregates. J. Ultrastruct. Res., 22, 303–311, 5 Figs, 1 Table, 1968.

Böhmers, J. C. A.: Bau und Struktur von Schale und Sipho bei permischen Ammonoideen. Amsterdam Univ. Geol. Inst., Med. 66, 1936.

Bøggild, O. B.: The shell structure of the mollusks. Kongl. Danske Vidensk. Selsk. Skr., Raekke 9, 2 (2), 233–326, 10 Textfigs., 15 Plates, København 1930.

Boyde, A., Bailey, E., Jones, S. J. and Tamarin, A.: Dimensional changes during specimen preparation for scanning electron microscopy. In: Proc. Tenth Annual Scanning Electron Microscope Symposium, 1, IITRI, Chicago, Illinois, 507–18 and 580, 1977 (quoted by Billings-Gagliardi and al.).

Bradley, D. E.: Replicas and Shadowing Techniques. In: Techniques for Electron Microscopy, ed. Desmond H. Kay, 2nd edition, Blackwell Sc. Publ., Oxford, p. 96–151, 24 Figs, 4 Tables, 1965.

Brown, Ph. R.: Pyritization in some Molluscan Shells. J. Sedim. Petrol., 36 (4), 1149–1151, 4 Figs, 1966.

Cantaluppi, G. and Rossi, M.: Cose e modalita di degradazione delle proteine nel tempo. Boll. Soc. Pal. Ital., 15 (1), 25–34, 1976.

Carlisle, D. B.: Chitin in a Cambrian fossil *Hyolithellus*. Biochem. J1, 90, 1C–2C, 1964.

Carriker, M. R.: Ultrastructural analysis of Dissolution of shell of the Bivalve *Mytilus edulis* by the Accessory Boring Organ of the Gastropod *Urosalpinx cinerea*. Marine Biology, 48, 105–134, 13 Plates, 79 Figs, 1978.

Chave, K. E.: Skeletal Durability and Preservation. In: Approaches to Paleoecology (eds. J. Imbrie and N. D. Newell), (John Wiley and Sons, Inc.), 377–387, 3 Figs, 2 Tables, New York, 1964.

Clausen, C. D.: Oberdevonische Cephalopoden aus dem Rheinischen Schiefergebirge. I. Orthocerida, Bactritida. Palaeontographica, 128, Abt. A, 1–86, 36 Textfigs, 10 Plates, Stuttgart 1968.

Cloud, P. E. Jr. and Barnes, V. E.: Early Ordovician Sea in Central Texas. In: Treatise on Marine Ecology and Paleoecology. Geological Society of America, H. S. Ladd, ed., Memoir 67, 2, 163, New York 1957.

Cornish, V. and Kendall, P. F. On the mineralogical constitution of calcareous organisms. Geol. Magaz., 5, 66–73, 1888.

Crenshaw, M. A.: The soluble matrix from *Mercenaria mercenaria* shell. Biomineralisation Forschungsber., 6, 5–11, 6 Figs, 1972.

— and Ristedt, H.: Histochemical and Structural Study of Nautiloid Septal Nacre. Biomineralisation Forschungsber., 8, 1–5, 3 Plates, 1975.

Currey, J. D.: Further studies on the mechanical properties of mollusc shell material. J. Zool. (London), 180, 445–453, 5 Textfigs, 1976.

— and Taylor, J. D.: The mechanical behaviour of some molluscan hard tissues. J. Zool. (London), 173, 395–406, 2 Tables, 3 Figs, 1 Plate, 1974.

Degens, E. T.: Diagenesis of organic matter. In: Diagenesis in Sediments. Developments in Sedimentology, 8 (eds Larsen Gunnar and G. V. Chilingar), Elsevier, Amsterdam 343–390, 1967.

De Jong, E. W., Westbroek, P., Westbroek, J. F. and Bruning, J. W.: Preservation of antigenic properties of macromolecules over 70 Myr. Nature, 252, 63–64, 1974

Dodd, J. R.: Processes of conversion of aragonite to calcite with examples from the Cretaceous of Texas. J. Sedim. Petrol., 36 733–741, 11 Figs, 1966.

Drozdova, T. V.: Geochemistry of amino acids and carbohydrates in old and recent sediments. In: Advances in Organic Geochemistry, 1973, Ed. Technip, 285–291, 2 Figs, 1973.

Erben, H. K.: Über die Bildung und das Wachstum von Perlmutt. Biomineralisation Forschungsber., 4, 15–46, 5 Textfigs, 6 Plates, 1972.

— On the structure and growth of the nacreous tablets in Gastropods. Biomineralisation Forschungsber., 7, 14–27, 5 Plates, 1974.

—, Flajs, G. und Siehl, A.: Die frühontogenische Entwicklung der Schalenstruktur ectocochleater Cephalopoden. Palaeontographica, 132, Abteilung A, 1–54, 12 Textfigs, 3 Tables, 15 Plates, Stuttgart 1969.

— and Reid, R. E. H.: Ultrastructure of shell, origin of conellae and siphuncular membranes in an ammonite. Biomineralisation Forschungsber., 3, 22–29, 2 Textfigs, 2 Plates, 1971.

Florkin, M. Grégoire, Ch., Bricteux-Grégoire, S. and Schoffeniels, E.: Conchiolines de nacres fossiles. C. R. Acad Sc. Paris, 252, 440–442, 1 Plate, 1961.

Flower, R. H.: Cherry Valley Cephalopods. Bull. Amer. Paleont, 22 (76), 273–366, 9 Plates, 1936.

— Study of the Pseudorthoceratidae. Palaeontogr. Amer., II (10) 1–214 and 245–460, 22 Textfigs, 9 Plates, 1939.

— Studies of the Actinoceratida. I. The Ordovician Development of the Actinoceratida with notes on actinoceroid Morphology and Ordovician Stratigraphy. New Mexico Bureau of Mines and Mineral Resources, Socorro, New Mexico, Mem. 2, 1–59, 5 Textfigs, 12 Plates, 1957.

— Nautiloid Shell Morphology. New Mexico Inst. Mineral. and Technol. State Bur. Mines and Mineral Res., Socorro, New Mexico, Mem. 13, 1–79; 23 Textfigs, 6 Plates, 1964.

— and Teichert, C.: The Cephalopod Order Discosorida. Univ. Kansas Paleontol. Contrib., Art. 6, 1–144, 35 Textfigs, 43 Plates, 1957.

Füchtbauer, H. and Goldschmidt, H.: Aragonitische Lumachellen im bitüminösen Wealden des Emslandes. Beitr. Miner. Petrogr., 10, 184–197, 8 Figs, 1964.

Goffinet, G.: Conchioline, Nacroïne et Chitine dans la coquille des Mollusques. Thesis in Zoology, University of Liège, Belgium, 1965.

— Etude au microscope électronique de structures organisées des constituants de la conchioline de nacre de *Nautilus macromphalus* Sowerby. Compar. Bioch. Physiol., 29, 835–839, 2 Figs, 1969.

— et Jeuniaux, Ch.: Composition chimique de la fraction "nacroïne" de la conchioline de nacre de *Nautilus pompilius* Linné. Compar. Biochem. Physiol., 29, 277–282, 3 Tables, 1969.

—, Grégoire, Ch. and Voss-Foucart, M. F.: On ultrastructure of the trabeculae in the interlamellar membranes of nacre conchiolin of the *Nautilus* shell. Arch. internat. Physiol. Bioch., 85 (5), 849–863, 4 Plates, 1977.

Gordon Jr, Mackenzie: Carboniferous Cephalopods of Arkansas. Geol. Survey U.S.A. Professional Paper 460, 322 pages, 30 Plates. U. S. Gov. Printing Office. Washington, 1964.

Grandjean, J., Grégoire, Ch. and Lutts, A.: On the mineral components and the remnants of organic structures in shells of fossil molluscs. Bull. Acad. Roy. Belg., Cl. des Sciences, 5e série, 50, 562–595, 5 Tables, 7 Plates, 1964.

Grégoire, Ch.: Topography of the organic components in mother-of-pearl. J. Biophys Biochem. Cytol., 3, 797–808, 1 Textfig., 7 Plates, 1957.

— Essai de détection au microscope électronique des dentelles organiques dans les nacres fossiles (ammonites, nautiloïdes, gastéropodes et pélécypodes. Arch. intern. Physiol. Biochim., 66, 674–676, 1958.

— A study on the remains of organic components in fossil mother-of-pearl. Bull. Inst. roy. Sc. natur. Belg., 35 (13), 1–14, 2 Tables, 8 Plates, 1959a.

— Conchiolin remnants in mother-of-pearl from fossil Cephalopoda. Nature, 184, 1157–1158, 3 Figs, 1959b.

— Further studies on structure of the organic components in mother-of-pearl, especially in Pelecypods. Part. I, Bull. Inst. roy. Sci. nat. Belg. 36 (23) 1–22, 1 Table, 5 Plates, 1960.

— On submicroscopic structure of the *Nautilus* shell. Bull. Inst. roy. Sc. natur. Belg., 38 (49), 1–71, 8 Textfigs, 2 Tables, 24 Plates, 1962.

— Thermal changes in the *Nautilus* shell. Nature, London 203, 868–869, 3 Figs, 1964.

— On organic remains in shells of Paleozoic and Mesozoic Cephalopods. (Nautiloids and Ammonoids). Bull. Inst. roy. Sci. natur. Belg., 42, (39), 1–36, 24 Plates, 1966a.

— Experimental Diagenesis in the Nautilus shell. In: Advances in Organic Geochemistry, (eds. G. D. Hobson and G. C. Speers), Pergamon Press, 429–442, 12 Figs, Oxford, London, 1966b.

— Sur la structure des matrices organiques des coquilles de mollusques. Biol. Rev., 42, 653–688, 1 Textfig., 6 Plates, 1967.

— Experimental alteration of the *Nautilus* shell by factors involved in Diagenesis and Metamorphism. Part. I. Thermal changes in conchiolin matrix of mother-of-pearl. Bull. Inst. roy. Sc. nat. Belg., 44 (25), 1–69, 2 Tables, 26 Plates, 1968.

— Structure of molluscan shells. In: Chemical Zoology, (eds. M. Florkin et B. T. Scheer), Acad. Press, Inc., New York, VII, 45–102, 2 Tables, 24 Figs, 1972a.

— Experimental alteration of the *Nautilus* shell by factors involved in Diagenesis and in Metamorphism. Part III. Thermal and hydrothermal changes in the organic and mineral components of the mural mother-of-pearl. Bull. Inst. roy Sc. natur., Belg., 48 (6), 1–85, 1 Table, 42 Plates, 1972b.

— Ultrastructure des composants organiques des coquilles de Mollusques. Haliotis, 2 (2), 51–79, 5 Plates, 1972c.

- On the submicroscopic structure of the organic components of the siphon in the *Nautilus* shell. Arch. Internat. Physiol. Bioch., 81 (2), 1 Textfig., 14 Plates, 1973.
- Unpublished observations (including compaction experiments) 1959–1979.
- , Duchâteau, Gh. und Florkin, M.: La trame protidique des nacres et des perles. Ann. Inst. Océanogr. Monaco, 31, 1–36, 8 Tables, 1 Textfig., 23 Plates, 1955.
- and Teichert, C.: Conchiolin membranes in shell and cameral deposits of Pennsylvanian cephalopods, Oklahoma. Okla. Geol. Notes, 25, 175–201, 1 Textfig., 11 Plates, 1965.
- , Gisbourne, Chr. M. and Hardy, A.: Über experimentelle Diagenese der Nautilusschale. Beitr. elektronenmikroskop. Direktabb. Oberfl., 2: 223–238, 13 Figs, 1969.
- and Voss-Foucart, M. F.: Proteins in shells of fossil Cephalopods (Nautiloids and Ammonoids) and experimental simulation of their alterations. Arch. Internat. Physiol. Bioch., 78, 191–203, 4 Textfigs, 16 Plates, 1970.
- and Lorent, R.: Alterations in conchiolin matrices of mother-of-pearl during conversion of aragonite into calcite under experimental conditions of pyrolysis and pressure. Biomineralisation Forschungsber., 6, 70–83, 8 Plates, 1972.
- , Goffinet, G. and Voss-Foucart, M. F.: Fixatives, Decalcifiers and Ultrastructure of the Conchiolin Remnants from mural nacreous Layers of fossil Cephalopod shells. Biomineralisation Forschungsber., 3 Plates (in the press).

Gümbel, J. von: Über die Beschaffenheit der Molluskenschalen. Zeitschr. Deutsch. Geol. Gesellsch., 36, 386, 1884.

Hall, A. and Kennedy, W. J.: Aragonite in fossils. Proc. Roy., Soc. B, 168, 377–412, 1 Table, 1967.

Hallam, A. and O'Hara, M.J.: Aragonitic fossils in the Lower Carboniferous of Scotland. Nature, London, 195, 273–274, 1962.

Hare, P. E.: The Amino Acid Composition of the Organic Matrix of some Recent and Fossil shells of some West Coast Species of *Mytilus*. Ph. D. Thesis, California Institute of Technology, Div. Geol. Sc., Pasadena, California, 1962.
- Geochemistry of proteins, peptides and amino acids. In: Organic Geochemistry, eds. G. Eglinton and M. T. J. Murphy, Springer Verlag, Berlin 438–463, 5 Figs, 4 Tables, 1969.
- Amino Acid dating of bone – The influence of water. Carnegie Institution of Washington, Yearbook 73, 576–581, 1 Table, 2 Figs, 1974.
- and Abelson, P. H.: Racemization of amino acids in fossil shells. Carnegie Institution of Washington Yearbook 66, 526–528, 2 Tables, 1967.
- and Mitterer, R. M.: Laboratory simulation of amino acid diagenesis in fossils. Carnegie Institution of Washington Yearbook 67, 205–208, 1 Table, 1 Fig., 1969.
- and Hoering, T. C.: The organic constituents of fossil mollusc shells. Carnegie Institution of Washington. Yearbook 76, 625–631, 3 Textfigs, 4 Tables, 1977.

Ho, Tong Yun: Stratigraphic and Paleoecologic applications of water-insoluble fraction of residual shell-proteins in fossil shells. Geol. Soc. America Bull., 77, 375–392, 4 Figs, 1966.

Hölder, H.: Über Gehäusebau, insbesondere Hohlkiel jurassischer Ammoniten. Palaeontographica A, 102, 18–48, 28 Textfigs, 5 Plates, Stuttgart 1952.
- und Mosebach, R.: Die Conellen auf Ammonitensteinkernen als Schalenrelikte fossiler Cephalopoden. N. Jhrb. Geol. Paläont, 92B (Abh.), 367–414, 25 Textfigs, 3 Tables, 1950.

Howarth, M. K.: The shell structure of the Liassic ammonite family *Dactylioceratidae*. British Museum (Natural History), Geology, Bull., 26, 45–67, 2 Textfigs, 10 Plates, 1975.

Hudson, J. D.: Pseudo-pleochroic calcite in recrystallized shell-limestones. Geol. Magaz., 99, 492–500, 1 Table, 1 Plate, 1962.
- The elemental composition of the organic fraction, and the water content, of some Recent and Fossil mollusc shells. Geochim. Cosmochim. Acta, 31, 2361–2378, 1967.

- The microstructure and mineralogy of the shell of a Jurassic Mytilid (Bivalvia). - Paleontol., 11 (2), 163–182, 5 Textfigs, 1 Table, 5 Plates, 1968.
- Pyrite in Ammonite Shells and in Shales. N. Jhrb. Geol. Paläontol., Abh., 157, 190–193, 1978.

Hyatt, A.: Fossil Cephalopods of the Museum of Comparative Zoology. Embryology. Bull. Mus. Compar. Zool., Harvard, 3, 59–111, 1872.

Iwata, K.: Ultrastructure of the conchiolin matrices in molluscan nacreous layer. Jl Fac. Sc., Hokkaido Univ., Ser. IV, 17(1), 173–229, 19 Plates, 1975a.
- An Application of Chromium Sulphate Demineralization to an Invertebrate calcified Tissue (Nacreous layer in Nautilus shell). Res. Rep. Fossil Club, 10, 6–11, 2 Plates, 1975b (in Japanese).
- Ulstrastructural Disintegration Trend of Fossil. Conchiolin. Jl Geol. Soc. Japan, 81, 155–164, 5 Plates, 1975c.

Jackson, T. A. and Bischoff, J. L.: The influence of amino acids on the kinetics of the recrystallization of aragonite to calcite. J. Geol., 79, 493–497, 1 Fig., 1971.

Kamiya, H.: Study on the Diagenetic and Experimental Alteration of some Aragonitic Shells. Basic Researches concerning the Early Process of their Fossilization. The Sc. Rep. of the Tokyo Kyoiku Daigaku, Section C (Geography, Geology and Mineralogy), 12, (118), 177–211, 19 Textfigs, 6 Tables, 10 Plates, 1975.

Kennedy, W. J.: The correlation of the Lower Chalk of South-East England. Proc. Geol. Assoc. England, 80, 459–551, 1969.
- and Hall, A.: The influence of organic matter on the preservation of aragonite in fossils. Proc. geol. Soc. London (n° 1643), 253–255, 1967.

Kitano, Y. and Hood, D. W.: The influence of organic material on the polymorphic crystallization of calcium carbonate. Geochim. Cosmochim. Acta, 29, 29–41, 5 Figs., 1965.
–, Kanamori, N. and Tokuyama, A.: Effects of organic matter on solubilities and crystal form of carbonates. Am. Zoologist, 9, 681, 1969.

Kummel, B.: Post-Triassic nautiloid genera. Bull. Mus. Comp. Zool., Harvard, 114 (7), 321–494, 28 Plates, 1956.

Land, Lynton, S. Diagenesis of Skeletal Carbonates. J. Sedim. Petrol., 37 (3), 914–930, 15 Figs, 1967.

Lowenstam, H. A.: Biologic problems relating to the composition and diagenesis of sediments. In: Donnelly, T. W., ed., The Earth Sciences: Univ. Chicago Press, Rice Univ. Semi-centenn. Publ., 137–195, 1 Table, 10 Figs, Chicago, 1963.

Macdonald, G. J. F. Experimental determination of calcite-aragonite equilibrium relations at elevated temperatures and pressures. Amer. Miner., 4, 744–756, 2 Figs, 3 Tables, 1956.

Masuda, Y., Sakurai, Y. and Osumi, M.: Shape of poly β benzyl-L-aspartate aggregate. J. El. Micr. (Japan) 25, 295–296, 1976.
- Spherical Form of Polypeptide Aggregate. J. El. Micr. (Japan) 27 (2), 145–146, 4 Figs, 1978.

Matter, Ph. III, Davidson, F. D. and Wyckoff, R. W. G.: The composition of fossil oyster shell proteins. Proc. Natl. Acad. Sc. Washington, 64, 970–972, 1969.

Mayer, F. K.: Über die Modifikation des Kalciumkarbonats in Schalen und Skeletten rezenter und fossiler Organismen. Chemie der Erde, 7, 346–350, 1932.

Miller, A. K. and Furnish, W. M.: Studies of Paleozoic Ammonoids. Middle Pennsylvanian *Schistoceratidae* (Ammonoidea). J. Pal., 32 (2), 253–268. 9 Textfigs, 2 Plates, 1958.
–, – and Schindewolf, O. H.: Paleozoic Ammonoidea. In: Treatise on Invertebrate Paleontology ed. R. C. Moore, Part L, Mollusca 4, Cephalopoda-Ammonoidea, L11–L 79, Figs, 1–123. Geol. Soc. of America and Univ. of Kansas Press, Lawrence, 1957.

Mitterer, R. M.: Amino Acid and Protein Geochemistry in Mollusk Shell. Ph. D. Thesis, The Florida State University, Geology, 151 pages, 27 Tables, 1966.

Moore, R. C.: Treatise on Invertebrate Paleontology Part L, Mollusca 4, Cephalopoda-Ammonoidea. 490 pages, 558 Figs, Geol. Soc. of America and University of Kansas Press, 1957.

— Treatise on Invertebrate Paleontology, Part K. Mollusca 3. Cephalopoda. Nautiloidea. 519 pages, 361 Figs, Geol. Soc. of America and Univ. of Kansas Press, Lawrence, 1964.

—, Lalicker, C. G. and Fischer, A. G.: Invertebrate Fossils.-Mac Graw Hill Book Co, Inc., 776p., 1952.

Murphy, James L.: The Pennsylvanian Nautiloid *Kionoceras ungeri* (Sturgeon and Miller). J. Paleont., 40 (6), 1388–1390, 3 Figs, 1966.

— Coiled Nautiloid Cephalopods from the Brush Creek Limestone (Conemaugh) of Eastern Ohio and Western Pennsylvania. J. Paleont., 44 (2), 195–205, 1 Textfig., 2 Plates, 1970.

Mutvei, H.: On the shells of *Nautilus* and *Spirula* with notes on the shell secretion in non-cephalopod molluscs. Arkiv. Zool., 16, 221–278, 30 Textfigs, 22 Plates, 1964.

— On the microscopic shell structure in some Jurassic ammonoids. N. Jhrb. Geol. Paläont., Abh., 129, 157–166, 4 Textfigs, 1 Plate, 1967.

— On the micro- and ultrastructure of the conchiolin in the nacreous layer of some recent and fossil molluscs. Stockholm Contrib. Geol., 20,1–17, 2 Textfigs, 17 Plates, 1969.

— Ultrastructure of the mineral and organic components of molluscan nacreous layers, Biomineralisation Forschungsber., 2, 49–72, 5 Textfigs, 11 Plates, 1970.

— Ultrastructural studies on cephalopod shells: Part I: The septa and siphonal tube in *Nautilus*. Bull. Geol. Inst. Univ. Uppsala. 3, 237–261, 11 Textfigs, 25 Plates, 1972a.

— Ultrastructural studies on Cephalopod shells: Part II: Orthoconic Cephalopods from the Pennsylvanian Buckhorn asphalt. Bull. Geol. Inst. Univ. Uppsala, New Series, 3, 263–272, 1 Textfig., 12 Plates, 1972b.

— Ulstrastructural relationships between the prismatic and nacreous layers in *Nautilus* (Cephalopoda). Biomineralisation, Forschungsber., 4, 80–86, 2 Figs, 2 Tables, 1972c.

— The nacreous layer in *Mytilus, Nucula* and *Unio*. (Bivalvia). Calcified Tissue Res., 24, 11–18, 21 Figs, 1977.

Newell, N. D.: Status of Invertebrate Paleontology 1953, V. Mollusca Bull. Mus. Comp. Zool, Harvard, 112, 161–172, 1954.

Nathusius-Königsborn, W. von: Untersuchungen über nichtcelluläre Organismen, namentlich Crustacean-panzer, Molluskenschalen und Eihüllen, 144 p. Berlin 1877 (quoted by Biedermann, 1902, 1914 and Schmidt, 1923).

Olsson, A. A.: Contributions to the Tertiary Paleontology of northern Peru: Part I: Eocene Mollusca and Brachiopoda. Bull. Amer. Paleont., 14, 1–154, 1928.

Palframan, D. F. B.: Mode of early shell growth in the ammonite *Promicroceras marstonense* Spath. Nature, London, 216, 1128–1130, 3 Figs, 1967.

Pettijohn, F. J.: Lithification and Diagenesis. In: Sedimentary Rocks (Second Edition, Pettijohn, F. J., ed., Harper & Bros, New York), Chapter 14, 648–681, 1957.

Pollard, E. C.: Thermal effects on protein, nucleic acid and viruses. In: The Structures and Properties of Biomolecules and Biological systems, Advances in Chemical Physics, VII, 201–237, J. Duchesne, ed., Inter Sciences Publishers London, New York and Sydney, 1964.

Quenstedt, W.: Über Erhaltungszustände von Muscheln und ihre Entstehung. Palaeontographica, 71, 1–65, Stuttgart 1929.

Reyment, R. A. and Eckstrand, O. R.: X-ray determinations on some cephalopod shells. Stockholm Contrib. Geol., 1, 91–96, 1 Table, 1957.

Römer, O.: Untersuchungen über den feineren Bau einiger Muschelschalen. Z. Wiss. Zool., 75, 437–472, 1903.

Ristedt, H.: Zum Bau der Orthoceriden Cephalopoden. Palaeontographica 137A, 155–195, 5 Tables, 7 Textfigs, 15 Plates, Stuttgart 1971.

Sacchi-Vialli, G.: Le sostanze organiche nei fossili. Atti Ist. Geol. Univ. Pavia, 14, 20–68, 3 Tables, 1963.

Samata, T.: Paläobiochemische Untersuchungen der löslichen Fraktion in den organischen Matrices von Pelecypoden-Schalen. Dr. Inaug. Dissert., University of Bonn, 208 pages, 60 Figs, 25 Tables, 1979.

Saunders, W. B.: Upper Mississippian Ammonoids from Arkansas and Oklahoma. Geol. Soc. Amer., Special Paper 145, 1–110, 1973.

Schmidt, W. J.: Bau und Bildung der Perlmuttermasse. Zool. Jhrb. (Abt. Anat); 45, 1–148, 5 Plates, 1923.

— Die Bausteine des Tierkörpers in polarisiertem Lichte. – Fr. Cohen, ed., Bonn, 528 p. 230 Figs, 1924.

— Perlmutter und Perlen. In: Die Rohstoffe des Tierreichs, ed. F. Pax und W. Arndt, Band II, 122–160, Berlin 1928.

Schoute-Vanneck, C. A.: A chemical aid for relative dating of coastal shell middens. South African J. Sc., 67–70, 1960.

Stehli, F. G.: Shell Mineralogy in Paleozoic Invertebrates. Science, 123 (N° 3206), 1031–1032, Washington D. C., 1956.

Stenzel, H. B.: Living Nautilus. In: Treatise on Invertebrate Paleontology, Part K, Mollusca 3, Cephalopoda-Nautiloidea, R. C. Moore, ed. Geol. Soc. Amer. and University of Kansas Press, K 59–K 93, Figs 43–68, Lawrence 1964.

Sundström, B.: Histological Decalcification using aqueous Solutions of Basic Chromium (III) Sulphate. Odont. Revy (Lund), 19, 3–19, 1968.

— and Zelander, T.: Routine Decalcification of thin sawed Sections of adult human Enamel by means of a basic Chromium (III) Sulphate Solution of stable ph.-Odont. Revy (Lund), 19, 234–263, 1968.

Switzer, G. and Boucot, A. J.: The mineral composition of some microfossils, J. Paleont, 29, 525–533, 1955.

Tasch, P.: Paleoecologic observations on the orthocerid coquina beds of the Maquoketa at Graf, Iowa. J. Paleont. 29 (3), 510–518, 8 Figs, 1955.

Taylor, J. H.: Some aspects of diagenesis. Advanc. Science, 20, 417–436, 7 Figs, 2 Tables, 1964.

Taylor, J. D. and Layman, M.: The mechanical properties of Bivalve (Mollusca) shell structures. Palaeont., 15, Part 1, 73–87, 10 Textfigs, 5 Tables, 1972.

Tazaki, K.: Nitrogen Analysis by micro-Kjeldahl method of the fossil shells yielded from the Johmon shell-mounds of the Kanto District, Japan. Earth Science (Chikyu Kagaku), 21, 21–24, 1967 (quoted by Kamiya, 1975).

Teichert, C.: Actinoceratoidea. In: Treatise on Invertebrate Paleontology. Part K Mollusca 3, K 190–K 216, Figs 125–152. The Geol. Soc. of America and The Univ. of Kansas Press, Lawrence, 1964.

— Major Features of Cephalopod Evolution. In: Essays in Paleontology and Stratigraphy, Raymond C. Moore Commemorative Volume, University of Kansas, Department of Geology Special Publication, 2, 163–210, 20 Figs, 1967.

Totten, D. K., Jr., Davidson, F. D. and Wyckhoff, R. W. G.: Amino acid composition of heated oyster shells. Proc. Natl. Acad. Sc. U.S.A., 69, 784–785, 2 Tables, 1973.

Towe, K. M. and Cifelli, R.: Wall Ultrastructure in the calcareous Foraminifera: Crystallographic aspects and a model for calcification. J. Paleont., 41, 742–762, 12 Plates, 1967.

— and Hamilton, G. H.: Ultrastructure and Inferred Calcification of the Mature and Developing Nacre in Bivalve Mollusks. Calc. Tiss. Res., 1, 306–318, 16 Figs, 1968.

— and Thompson, G. R.: The structure of some Bivalve shell Carbonates prepared by Ion – Beam Thinning. A comparison study. Calc. Tiss. Res. 10, 38–48, 4 Figs, 1972.

Turekian, K. K. and Armstrong, R. L.: Chemical and mineralogical composition of fossil molluscan shells from the Fox Hill Formation, South Dakota (Cretaceous). Bull. Geol. Soc. Amer., 72, 1817–1828, 2 Figs, 8 Tables, 1961.

Unklesbay, A. G.: Pennsylvanian Cephalopods of Oklahoma. Oklahoma Geol. Survey Bull., 96, 1–150, 16 Textfigs, 19 Plates, 1962.

Uozumi, S. and Iwata, K.: Studies on calcified tissues. Part. II. Comparison of ultrastructure of the organic matrix region of recent and fossil *Mytilus*. J. Geol. Soc. Japan, 75, 417–424, 5 Plates, 1969.

Vallentyne, J. R.: Biochemistry of organic matter. – II. Thermal reaction kinetics and transformation products of amino compounds. Geoch. Cosmoch Acta, 28, 157–188, 8 Tables, 8 Figs, 1964.

— Pyrolysis of amino acids in Pleistocene *Mercenaria* shells. Geoch. Cosmoch. Acta, 33, 1453–1458, 1 Fig., 2 Tables, 1969.

Voss-Foucart, M. F.: Constituants organiques de la coquille des Céphalopodes et phénomènes de Paléisation. Ph. D. Thesis, Liège University, Belgium, 1970.

—, Laurent, Cl. and Grégoire, Ch.: Sur les constituants organiques des coquilles d'Ethérides. Arch. intern. Physiol. Bioch., 77 (5), 901–915, 3 Figs, 1 Plate, 1969.

— and Grégoire, Ch.: Biochemical composition and submicroscopic structure of matrices of nacreous conchiolin in fossil Cephalopods (Nautiloids and Ammonoids). Bull. Inst. roy. Sc. natur. Belg., 47 (41), 1–43, 4 Tables, 11 Plates, 1971.

— On biochemical and structural alterations in fossil and in pyrolyzed, modern mother-of-pearl. Biomineralisation Forschungsber., 6, 134–140, 4 Textfigs, 3 Plates, 1972.

— Experimental alteration of the *Nautilus* shell by factors of diagenesis and metamorphism. II. Amino acid patterns in the conchiolin matrix of the pyrolyzed, modern mother-of-pearl. Bull. Inst. roy, Sc. natur, Belg. 49 (9), 1–13, 7 Figs, 2 Tables, 3 Plates, 1973a.

— Conchioline de Céphalopodes fossiles et simulation de ses altérations. Adv. in Org. Geochemistry, 1973, 279–284, 4 Figs, 1973b.

— On Biochemical and Structural Alterations of the Nacre Conchiolin in the *Nautilus* Shell under Conditions of protracted, moderate Heating and Pressure. Arch. internat. Physiol. Bioch., 83, 43–52, 1 Table, 12 Figs, 1975.

Wada, K.: Crystal growth of molluscan shells. – Bull. Natl. Pearl Res. Labor., Kashikojima, 7, 703–828, 20 Textfigs, 149 Figs, 1961.

Watabe, N.: Studies on shell formation. XI Crystal-matrix relationships in the inner layers of mollusk shells. J. Ultrastr. Res., 12, 351–370, 1965.

— and Dunkelberger, V. G.: Ultrastructural studies on calcification in various organisms. In: Scanning Electron Microscopy, SEM, Inc., AMF O'Hare, III. 60666, U.S.A., 11, 403–416, 23 Figs, 1979.

Weiner, S. and Hood, L.: Soluble protein of the organic matrix of mollusk shells: a potential template for shell formation. Science (Washington), 190, 987–989, 2 Tables, 1 Fig., 1975.

—, Lowenstam, H. A. and Hood, L.: Characterization of 80-million-year-old mollusk shell proteins. Proc. Natl Acad. Sc. U.S.A. (Washington), 73 (8), 2541–2545, 2 Figs, 2 Tables, 1976.

—, —, Taborek, B. and Hood, L.: Fossil mollusk shell organic matrix components preserved for 80 million years. Paleobiology 5 (2), 144–150, 2 Figs., 2 Tables, 1979.

Westbroek, P., van der Meide, J. S., van der Wey-Kloppers, J. S., van der Sluis, R. J., de Leeuw, J. W. and de Jong, E. W.: Fossil macromolecules from Cephalopod shells: Characterization, immunological response and diagenesis. Paleobiology, 5 (2), 150–167, 6 Figs, 1979.

Williams, G. and Jackson, D. S.: Two organic fixatives for acid mucopolysaccharides. Stain Technol., 31, 189–191, 1956.

Wise, Jr. Sh. W.: Microarchitecture and mode of formation of nacre (mother-of-pearl) in Pelecypods, Gastropods and Cephalopods. Eclogae Geol. Helvet, 63, 775–797, 4 Textfigs, 1 Table, 10 Plates, 1970.

Wray, J. L. and Daniels, F.: Precipitation of calcite and aragonite. J. Amer. Chem. Soc., 79, 2031–2034, 1957.

Wyckoff, R. W. G.: The Biochemistry of animal fossils. 152 Pages, 39 Tables, 41 Figs, Scientechnica (Publishers) Ltd, Bristol 1972.

Yochelson, E. L., White Jr., J. S. and Gordon Jr., M.: Aragonite and calcite in mollusks from the Pennsylvanian Kendrick Shale (of Jillson) in Kentucky. U. S. Geol. Survey Professional Paper, 575-D, D-76-D78, 1 Table, 1967.

Zobell, Cl. E.: Organic Chemistry of Sulfur. In: Organic Geochemistry, Monogr. N° 16, Earth Sciences Series, ed. I. A. Breger, Sympos. Public. Divis., Chapter 13, 543–578, New York 1963.

8. ACKNOWLEDGEMENTS

I am deeply indebted to the persons, listed below, from the Departments of Geology and Paleontology of the following Institutions, for generous gifts of material: Prof. Andrea Allasinaz, University of Milano (1143, 1157); Prof. Lucia Baccelle-Scudeler, University of Ferrara (1156); Mr. S. Barney-Hansford, Charmouth, Dorset, England (910, 916, 921 to 952, 972, 973, 1160, 1161, 1174, 1179); Prof. Tove Birkelund, University of Copenhagen (363, 364, 365, 366, 1159); Prof. Joseph Bouckaert, Service Géologique de Belgique (663 to 669, 519; 713, 675, 1072, 1073, 1074); Prof. C. C. Branson, Oklahoma Geological Survey, Norman, Oklahoma, U.S.A. (530, 607, 749 to 755); The British Portland Cement Manufacturers Ltd (Mr. E. Hamilton Clements, Manager) in Smalldole, Steyning, Sussex, England (648 to 659); Dr. James Craig Brown, Little Rock, Arkansas, U.S.A. (1203 to 1221); Prof. Kenneth E. Caster and Prof. Kenneth A. Beem, University of Cincinnati, Ohio, U.S.A. (1030, 1031, 1032, 1050, 1051); Dr. R. Catala-Stucki, Aquarium de Nouméa, New Caledonia (fresh N*autilus* ma*cromphalu*s); Dr. W. A. Cobban, United States Geological Survey, Paleontology-Stratigraphy Branch, Federal Center, Denver, Colorado, U.S.A. (1002 to 1005; Mesozoic Localities D 2624, D 903, 23625, 21740); Dr. A. E. Cockbain, Geological Survey of Western Australia, Perth, W. A. (1052, 1064 to 1069, F 5234, 19951); Dr. G. Arthur Cooper, United States National Museum, Washington D.C., U.S.A. (386 to 400, 608); Dr. Thomas A. Darragh, National Museum of Victoria, Melbourne, Australia (1076 to 1081); Prof. Heinrich Karl Erben, University of Bonn, Germany (625); Prof. A. G. Fischer, University of Princeton, U.S.A. (530, 421); Prof. Rousseau H. Flower, New Mexico Institute of Mining and Technology, Socorro, New Mexico, U.S.A. (422, 453, 454, 697 to 705); Prof. C. L. Forbes, Sedgwick Museum, Cambridge, England (535 to 540); Dr. J. L. Freeman and Mr. V. Hackney, The London Brick Company Ltd, Peterborough, England (465 to 475); Prof. William M. Furnish and Prof. Brian F. Glenister, State University of Iowa, Iowa City, Iowa (356 to 361, 676, 677, 720, 855 to 858, 768, 769); Prof. C. H. Holland, University of Dublin, Ireland (567); Dr. M. Glibert, Institut royal des Sciences naturelles de Belgique, (323, 333, 335); Prof. Mackenzie Gordon, United States Geological Survey and National Museum, Smithsonian Institution, Washington D.C., U.S.A. (969); Dr. Peter Jung, Naturhistorisches Museum, Basel, Switzerland (1251, 1252); Prof. Abbé Albert F. de Lapparent, Université Catholique de Paris (429, 430, 455 to 463); Prof. J. F. Kirkaldy, Queen Mary College, University of London (for information about the quarries of Smalldole); Dr. P. L. Maubeuge, St-Max, Meurthe et Moselle, France (339 to 345, 362, 367); Dr. Claude Monty, University of Liège (674); Prof. Raymond C. Moore†, State Geological Survey and University of Kansas (433, 434, 436, 10540, 11775); Prof. James L. Murphy, Case Western Reserve University, Cleveland, Ohio, U.S.A. (1130 to 1139); Prof. Norman D. Newell, American Museum of Natural History and Columbia University, New York City (385, 386); Dr. P. Rancurel and Dr. Yves Magnier, Office de la Recherche Scientifique et Technique, Centre de Nouméa, New Caledonia (fresh shell of *Nautilus macromphalus*); Dr. J. B. Reeside, The United States Geological Survey, Washington, D.C. (for authorizing gifts of samples); Prof. Dr. Rosalvina

Rivera, Directora del Museo Nacional de Paleontologia de la Universidad Nacional de Ingeniera, Lima, Peru (1119 to 1126) and Dr. Isolina Corrales Bravo (for selection of specimens); Miss Samuel, Curator, Museum of Dorchester, Dorset, England (923, 926); Prof. Giulia Sacchi-Vialli, University of Pavia, Italy (1025, 1026); Prof. O. H. Schindewolf†, University of Tübingen (402 to 409); Dr. Norman F. Sohl, The United States Geological Survey, Washington D.C. (387 USGS 18756; 388, 1005: USGS 21740, 489: USGS 20843; 394: USGS 19003; 398, 399: USGS 27065; 399: USGS 27065; 397: USGS 1882; 400: USGS 5879; 401: USGS 10771; 998: USGS 25406; 1001: USGS 17809 and 25483; 004: USGS 23625 and 1006); Prof. R. A. Reyment, University of Uppsala, Sweden (551); Prof. W. Bruce Saunders, Bryn Mawr College, Bryn Mawr, Pennsylvania, U.S.A. (shells and siphuncles from two freshly caught *Nautilus* sp.); Prof. Francis G. Stehli, Case Western Reserve University, Cleveland, Ohio, U.S.A. (420, 421); Prof. Curt Teichert, University of Rochester, Rochester, U.S.A. (902); Prof. A. G. Unklesbay (553); Dr. Thomas R. Waller, Department of Paleobiology, United States National Museum of Natural History, Smithsonian Institution, Washington D. C. (1253, 1254); Prof. Franz Westphal, University of Tübingen (785 to 799); Dr. R. B. Wilson, H. M. Geological Survey of Scotland, Edinburgh, Scotland (662); Dr. Hans K. Zöbelein, Bayerische Staatssammlung für Paläontologie und historische Geologie (555, 556, 557).

I also thank Dr. Michael K. Howarth, Curator of fossils, British Museum of Natural History, for identification of Jurassic ammonites of England, Mr. D. Phillips, for checking the determination of several specimens, Dr. J. Cl. Bussers, for applying the Anderson's critical point drying method to a part of the material. I thank Mrs Dr. M. F. Voss-Foucart for several suggestions about the text, Prof. H. K. Erben for reading the manuscript and presenting the paper to the Mainz Academy of Sciences. I also thank Dr. X. Misonne, editor of the Bulletin de l'Institut royal des Sciences naturelles de Belgique and Dr. Gh. Duchâteau, editor of the Archives internationales de Physiologie et de Biochimie, for authorizing reproduction of figures or parts of figures of former papers.

The research has been financed by the Fonds de la Recherche Fondamentale Collective and by the Fonds National de la Recherche Scientifique. The help of these Institutions is gratefully acknowledged.

I am deeply grateful to Prof. Dr. Pierre Coheur, Director, for his hospitality in the Centre de Recherches Métallurgiques from October 1948. I also thank the staff of the Department of Métallography of the CRM for maintaining the TEM and the SEM in excellent working conditions.

Centre de Recherches Métallurgiques (C. R. M.), Abbaye du Val Benoît, 4000 Liège, Belgium.
Department of General and Comparative Biochemistry, University of Liège.

9. EXPLANATION OF FIGURES

Plate 1.

Fig. 1: *Wellerites mohri* Plummer & Scott (Ammonoid) (769). Pennsylvanian, Buckhorn asphalt, Sulphur Oklahoma, U.S.A. Specimen given by Prof. William M. Furnish and Prof. Brian F. Glenister (S.U.I. 8891).
  Mineral composition of the samples: Aragonite and Calcite.
  Surface of transverse fracture of the shell wall. Direct replica (Carbon-Platinum).
  Preservation, in this region of the nacreous layer of the shell wall, of the original, mineral (stacks of aragonite tabular crystals) and conchiolin components (horizontal, interlamellar sheets (small asterisk) and vertical, intercrystalline matrices (large asterisk). The latter are obliquely involved in the surface of fracture. In these intercrystalline matrices, the conchiolin trabeculae appear in the form of parallel, tortuous, ribbon-shaped structures.
  TEM: × 36 000
  (From Grégoire, 1972b, Pl. III, Fig. 7.) Reproduced with permission of the Editor of the Bulletin de l'Institut royal des Sciences naturelles de Belgique.

Fig. 2: *Psiloceras planorbis* (J. de C. Sowerby) (988). (Ammonoidea *Psilocerataceae* Lower Jurassic, Lower Lias, Planorbis Zone Range, Hettangian, Somerset coast of England.
  Mineral composition of the sample: Aragonite.
  SEM micrograph of a surface of transverse fracture of a characteristic flattened shell, showing, from right to left, the inner part of the nacreous layer of the shell wall, the inner prismatic layer (helle Schicht: step-like structures) and the adjacent inner mould.
  × 1600.

Fig. 3: Unidentified nautiloid (Nautilida) (1124). Eocene, Lobitos, Piura, Peru. Specimen given by Prof. Rosalvina Rivera.
  Mineral composition of the sample: Calcite.
  SEM micrograph of a surface of transverse fracture of the shell wall, showing the complete inversion, rather infrequently found in Eocene nautiloids, of the original crystals of aragonite to euhedral crystals of calcite.
  × 100.

Figs 4 and 5: *Ammonites lineatus penicillatus* (785). (Ammonoidea, Lytocerataceae) Middle Jurassic, Brauner Jura, Gammelshausen bei Boll, Württemberg, Germany. Specimen given by Prof. O. H. Schindewolf, Prof. Ad. Seilacher and Prof. Fr. Westphal, Tübingen.
  Mineral composition of the samples: Aragonite and Calcite.

Transverse fracture of the shell wall. Direct replicas (carbon-platin). Figs 4 and 5 (TEM) show different stages of transformation (inversion) of aragonite to calcite.

In Fig. 4 two (asterisks) former columns or stacks of tabular crystals of aragonite, considerably altered, are seen running horizontally. In the upper stack, the following structures are visible from left to right: 1. an anhedral block, (asterisk), formed by coalescence of four tabular crystals of aragonite and in which the interlamellar spaces are still faintly visible; 2. a still isolated tabular crystal; 3. two tabular crystals fused together at their lower part; 4. another individual tabular crystal; 5. a row of aligned blocks with anhedral or subhedral, irregular shapes. In the lower stack, the tabular crystals, fused by coalescence, are no more recognizable.

Fig. 5 depicts a region of the fractured shell wall composed of parallel rows of coarse, subhedral (plane facets) blocks (asterisk). These alignments are the only traces of the stacking of the aragonite crystals in the original architecture of the shell wall.

TEM. Fig. 4: × 15 000; Fig. 5: × 9 000.

(Fragments of Figs 40 and 42, Pl. XIV, in Grégoire, 1972b). Reproduced with permission of the Editor of the Bulletin de l'Institut royal des Sciences naturelles de Belgique.

Plate 2.

Fig. 6: Ammonites *lineatus penicillatus* (785). Middle Jurassic. Brauner Jura, Gammelshausen bei Boll, Württemberg, Germany. Specimens given by Prof. O. H. Schindewolf, Ad. Seilacher and Fr. Westphal.

Mineral composition of the samples: Aragonite and Calcite.

Transverse fracture of the nacreous layer of the shell wall. Direct replica (carbon-platinum).

Surrounded by a mosaic of coarse anhedral or subhedral (extreme left) calcite crystals, relics of the original brickwall architecture of the nacreous layer of the shell wall (columnar stacking) (upper right part of the Fig.) show different stages in the process of inversion of aragonite to calcite, which occurs by coalescence of groups of contiguous aragonite crystals in the stacks. One of the blocks (asterisk) is formed by coalescence of three superimposed crystals.

TEM: × 27 000

Figs 7 and 8: *Dolorthoceras sociale* (Hall) (357–A2 and 677). Nautiloid. Orthocerid. Upper Ordovician, Lower Maquoketa Formation, Graf, Iowa, U.S.A. Specimen given by Prof. W. M. Furnish and Prof. B. F. Glenister.

Mineral composition of the samples: Carbonate apatite (Dahlite), Navajoite and SiO^2. Direct replicas (carbon-platinum) of the iridescent, golden-yellow material coating the inner, concave part of the shell wall.

Fig. 7: Cone-shaped sector of a spherulite. Fan-shaped arrangement of crystalline rods and laths radiating from a centre.

Reversed print (black shadows downwards).

TEM. × 24 000.

In Fig. 8, traces (in white. asterisk) of the organic sheaths appear in the form of faintly white ribbons around the crystals, on a slightly etched surface (see Fig. 58).

TEM: × 42 000

Plate 3.

Figs 9 and 10: *Nautilus pompilius* Linné (449–122) (modern). EDTA-insoluble fraction of conchiolin interlamellar matrix, collected from the inner part of the nacreous layer of the shell wall in the living chamber, immediately under the (discarded) inner prismatic layer.
  Mineral composition of the samples: aragonite.
  Robust, irregularly cylindrical trabeculae, studded with hemispheric protuberances of different sizes (see Textfig. 1) and a generally elongate fenestration, are the features of the nautiloid pattern of conchiolin. Trabeculae and fenestration slightly vary in shape in stretched or shrunk fragments of conchiolin sheets.
  Shadow-cast with palladium. Direct prints (white shadows upwards). TEM: × 48 000

Figs 11 and 12: *Nautilus pompilius* Linné (modern). Nacreous layer of the shell wall compressed at room temperature (20 °C) at 13 000 kgs (about 13 Kilobars) per square cm. for 30 minutes (DB 20–743) (Fig. 11) and at 50 000 kgs (about 50 kilobars) per square cm. for 5 hours (DB-4-740) (Fig. 12).
  Mineral composition of the samples: aragonite.
  EDTA – insoluble fractions of the conchiolin sheets. In both samples, the differences with the control samples (Figs. 9 and 10) about the form of the trabeculae and the outlines of fenestration do not show with certainty modifications, produced by pressure, of the nautiloid pattern. Constriction of the trabeculae and rounding up of fenestration, visible in Fig. 11, may be observed in the central regions of the polygonal fields of normal interlamellar membranes.
  Shadow-cast with palladium (Fig. 11) and platinum (Fig. 12). Direct prints.
  TEM: × 48 000.

Plate 4.

Fig. 13: Unidentified Orthocerid (Nautiloidea) (607-3). Pennsylvanian, Buckhorn Asphalt, Sulphur, Oklahoma, U.S.A. Specimen given by Prof. C. C. Branson.
  Mineral composition of the samples: Aragonite and ray 3.03 Å of calcite.
  With the exception of a diffuse flattening of the trabeculae and a decrease in the number of protuberances, the EDTA-insoluble fraction of interlamellar conchiolin from the nacreous layer of the shell wall shows preservation of the features of the nautiloid pattern in four polygonal fields delimited by protruding intercrystalline cords.
  Shadow-cast with platinum. Direct print.
  TEM: × 48 000
  (From Grégoire and Voss-Foucart, 1970, Pl. 4). Reproduced with permission of the Editor of the Archives internationales de Physiologie et de Biochimie.

Fig. 14: Crushed, unidentified ammonoid (*Gastrioceras?*) (720-7). Pennsylvanian, Buckhorn asphalt, Sulphur, Oklahoma, U.S.A. Specimen given by Prof. William M. Furnish and Prof. Brian F. Glenister.
  Mineral composition of the samples: Aragonite.
  Conchiolin of the shell wall. The pattern differs from the nautiloid, mural (see Fig. 13) and

septal (Fig. 15) patterns. As in conchiolin from nacreous layers of the shell wall of other ammonoids, the trabeculae are dumpy, the fenestration irregularly rounded.
Shadow-cast with platinum. Direct print.
TEM: × 50 000

Fig. 15: *Pseudorthoceras knoxense* Mc Chesney (Nautiloidea, Orthocerida) (421-9-78a), Pennsylvanian Buckhorn asphalt, Sulphur, Oklahoma, U.S.A. Specimen given by Prof. Fr. G. Stehli.
Mineral composition of the samples: Aragonite.
As in the modern *Nautilus,* the pattern of texture of the septal conchiolin is more slender than that of the mural conchiolin (see Fig. 13). Fragments of intercrystalline conchiolin, which separate the polygonal fields composing the interlamellar membranes of mother-of-pearl, are visible in the form of cords on the right part of the figure.
Shadow-cast with platinum. Direct print.
TEM: × 36 000.

Plate 5.

Fig. 16: Unidentified, gyroconic, loosely coiled nautiloid of large size (diameter of the outer whorl: 21 cm) (713-1367). Middle Devonian, Couvinian, Dinant, Belgium. Specimen given by Prof. J. Bouckaert.
Mineral composition of the samples: predominance of calcite.
Phase contrast micrograph of a cluster of interlamellar membranes of conchiolin from the nacreous layer of the shell wall. The cords of intercrystalline conchiolin which delimit the rounded polygonal fields of these membranes, appear in the form of sharp black lines. These clusters were stained in violet by the biuret reaction.
× 300.
(from Voss-Foucart and Grégoire, 1971, Fig. 15). Reproduced with permission of the Editor of the Bulletin de l'Institut royal des Sciences naturelles de Belgique.

Fig. 17: Unidentified Orthocerid (Nautiloidea) (607-6) (see Fig. 13). Specimen given by Prof. C. C. Branson.
Splinter of fracture of the shell wall, exfoliated into the internal mould. The EDTA-insoluble conchiolin matrix of this fragment shows the features of the modern mural nautiloid pattern.
Shadow-cast with palladium. Direct print.
TEM: × 20 000.

Fig. 18: *Rayonnoceras* sp. (1203-1a) Nautiloid. Actinocerida. Lower Carboniferous, Mississippian, Fayetteville Shale, Fayetteville, Arkansas, U.S.A. Specimen given by Dr. J. Craig Brown.
Mineral composition of the samples: predominantly α-quartz.
The substantial EDTA-insoluble conchiolin fraction from the nacreous layer of the shell wall consists in this sample of loose networks of mostly flattened, broadened angulate trabeculae (asterisk). This type of conchiolin alteration has been recorded in many conchiolin samples of all ages.
Shadow-cast with platinum. Direct print.
TEM: × 48 000.

Plate 6.

Fig. 19: *Rutoceratidae* sp. (Nautiloidea, Nautilida) (685-2) Upper Middle Devonian, Givetian, Sötenicher Mulde, Sötenich, Eifel, Germany. This gyroconic or cyrtoconic specimen, 3 cm in diameter, showed signs of multiple fractures and displacements of the wall and the septa.
 Mineral composition of the samples: Calcite.
 This EDTA-insoluble conchiolin network is composed of smooth, angulate and cylindrical trabeculae, which delimit a rounded polygonal fenestration.
 Shadow-cast with palladium. Reversed print.
 TEM: × 42 000.

Figs 20 and 21: *Endolobus clorensis* Collinson (969-71). Giant Nautiloidea, Nautilida. Carboniferous, Arkansas, U.S.A. Fragments given by Prof. Mackenzie Gordon (U.S.G.S. 15075 PC).
 Mineral composition of the samples: navajoite.
 EDTA-insoluble remnants o f the conchiolin matrix of the thick anthracite-black and grey nacreous layer of the shell wall. A network of sturdy, flattened or cylindrical trabeculae delimiting a rounded polygonal fenestration (Fig. 21,: centre; Fig. 20: bottom) is mixed with fragments of tortuous stumps, membraneous trabeculae and rounded particles.
 Shadow-cast with platinum. Direct print.
 TEM: 48 000.

Plate 7.

Figs 22 and 23: Unidentified Nautiloid of large size (diameter of the outer whorl: 30 cm). (Nautilida) 1123). Eocene, Lobitos, Piura, Peru. Specimen given by Prof. Rosalvina Rivera, Lima, Peru.
 Mineral composition of the samples, Calcite.
 EDTA-insoluble remnants of conchiolin from the nacreous layer of the shell wall.
 The alterations of the fossil conchiolin matrices (types 2 to 6: loose networks of broadened, angulate, flattened and cylindrical trabeculae, spheroidal particles, vesicles, membranes) do not distinctly differ in preparations dried in air (Fig. 23) and those treated by the critical point drying method (Fig. 22). In the present material, the method produced a general shrinkage of the structures (visible in Fig. 22) and multiple rounded perforations of the supporting formvar films (asterisk).
 Shadow-cast with platinum. Direct print.
 TEM: Figs 22 and 23: × 48 000.

Fig. 24: *Eutrephoceras* sp. (999) (Nautiloidea, Nautilida). Cretaceous, Eutaw Formation, Menabites, Alabama, U.S.A. Specimen given by Dr. Norman F. Sohl (U.S.G.S. 27065).
 Mineral composition of the samples: Aragonite.
 EDTA-insoluble remnants of conchiolin matrix of the nacreous layer of the shell wall, composed in this field of substantial, loose networks of flattened and swollen trabeculae, and of spheroidal particles.
 Shadow-cast with platinum. Direct print.
 TEM: × 48 000

Fig. 25: Nautiloid sp. (Nautilida: *Aturia?*) (1022). Eocene, Wemmelian, Baeleghem, Flanders, Belgium.

Mineral composition of the samples: Aragonite. In the abundant EDTA-insoluble conchiolin residues of the nacreous layer of the shell wall, the trabeculae of the loosened networks appear flattened into membranes, or cylindrical, varicose, with local hemispheric swellings The types of alteration do not distinctly differ from those observed in the conchiolin of the Cretaceous specimen shown in Fig. 24. In this specimen, the conchiolin structures displayed many vesicle-like particles of large size (asterisk) (see Voss-Foucart & Grégoire, 1971, 1972).

Shadow-cast with platinum. Direct print.

TEM: × 48 000

Plate 8.

Fig. 26: *Aturoidea distans* Teichert 1943 (1080-1). Palaeocene, Buckleys Point, S. W. of Princetown, Victoria, Australia. Specimen given by Dr. Thomas A. Darragh.

Mineral composition of the samples: Aragonite.

EDTA-insoluble conchiolin residues of the nacreous layer of the shell wall. The nautiloid pattern, with swollen trabeculae, is recognizable. An identical change has been previously recorded in the conchiolin residues of a Pennsylvanian ammonite from Buckhorn asphalt (Grandjean & al., 1964, Plate 2, Fig. 2).

Shadow-cast with platinum.

TEM: × 42 000. Direct print.

Fig. 27: *Aturia* sp. (926-1-71) (Nautiloidea, Nautilida). Palaeocene-Miocene (probably Eocene). Specimen given by Miss Samuel, Curator, Museum of Dorchester, England.

Mineral composition of the samples: Aragonite.

The EDTA-insoluble conchiolin residues of the nacreous layer of the shell wall appear in the form of elongate, flattened and broadened trabeculae and their fragments. Similar alterations, reproduced experimentally, are shown in Fig. 28.

Shadow-cast with platinum. Direct print.

TEM: × 48 000. Direct print.

Fig. 28: *Nautilus pompilius* Linné (modern) (1095 DB). Nacreous layer of the shell wall heated dry at 800 °C for 5 minutes under a pressure of 325 kgs par square cm.

Mineral composition of the samples: Aragonite.

The structural modifications developed in the conchiolin of this nacreous layer during a short exposure to heat and pressure consist of flattening and broadening of the trabeculae into varicose ribbons and nodules associated with hemispheric tuberosities. These tuberosities might be the altered protuberances visible on the original trabeculae of *Nautilus* (see Textfig. 1).

Shadow-cast with platinum. Reversed print.

TEM: × 48 000.

Plate 9.

Figs 29–32: Figs. 29–32 show in three Paleozoic Ammonoids the structural alterations in the EDTA-insoluble conchiolin residues of the nacreous layer of the shell wall. These residues appear predominantly in the form of loose networks of flattened, angulate trabeculae (Fig. 29: asterisk), powerful cords (Fig. 32) membranes produced by coalescence of the broadened trabeculae (Fig. 30: asterisk) and fragments of slender, cylindrical trabeculae, mixed with the other types of alteration (Fig. 31: asterisk).

Figs 29–30: *Tumulites varians* (1207-1a) (Ammonoidea, Goniatitina). Carboniferous, Upper Mississippian, Imo Shale, Saunders Locality, Arkansas, USA Specimen given by Dr. J. Craig Brown.
    Mineral composition of the samples: α-quartz.
    Shadow-cast with platinum. Direct print.
    TEM: Fig. 29: × 48 000. Fig. 30: × 60 000.

Fig. 31: *Cravenoceras fayettevillea* (1211). (Ammonoidea, Goniatitina). Carboniferous, Mississippian, Fayetteville, Shale, Fayetteville, Arkansas, U.S.A. Specimen given by Dr. J. Craig Brown.
    Mineral composition of the samples: Calcite.
    Shadow-cast with platinum. Direct print.
    TEM: 48 000.

Fig. 32: *Paraceltites elegans* Girty (Ammonoidea, *Otocerataceae*) (856). Late Lower Permian, Bone Spring Limestone, Culberson County, Texas, U.S.A. Specimen given by Prof. William M. Furnish and Prof. Brian F. Glenister.
    Mineral composition of the samples: Calcite.
    Shadow-cast with platinum. Direct print.
    TEM: × 48 000.

Plate 10.

Fig. 33: *Pronorites* sp. (754) (Ammonoidea, *Prolecanitina*). Pennsylvanian, Buckhorn asphalt, Sulphur, Oklahoma, U.S.A. Specimen given by Prof. C. C. Branson.
    Mineral composition of the samples: Aragonite and Calcite.
    In this specimen, the structure of the EDTA-insoluble conchiolin residues of the mural nacreous layer of the shell showed considerably flattened and broadened, angulate trabeculae, coalesced into membranes (see Textfig. 4b).
    Shadow-cast with platinum. Direct print.
    TEM: × 48 000.

Fig. 34: *Ceratitaceae* sp. (1026-2) (Ammonoidea, *Ceratitaceae*) Jurassic, Toarcian. Specimen given by Prof. G. Sacchi-Vialli.
    Mineral composition of the samples: Calcite.
    The EDTA-insoluble conchiolin fraction of the nacreous layer of the shell wall consists predominantly in this specimen of rounded flakes or slabs, probably produced during fossilization by fragmentation of membranes, themselves developed by coalescence of flattened trabeculae.
    Shadow-cast with platinum. Direct print.
    TEM: × 48 000.

Fig. 35: *Carnites floridus* Wulfen (965-Crs-3). (Ammonoidea, *Ceratitaceae*). Lumachelle of Ammonites and Mussel shells. Upper Triassic (Ceratiten Layers). Grube Ramser Oswaldi, Einbaulauf, Kraith bei Bleiberg, Carinthia, Austria.

Mineral composition of the samples: Aragonite.

Decalcification of the nacreous layer in chromium sulphate during 16 days. Flattened, cylindrical trabeculae and agglutinated spheroids are superimposed on to membranes (asterisk) produced by shrinkage and coalescence of networks of trabeculae, identical in structure to those found in EDTA-insoluble conchiolin material.

Shadow-cast with platinum. Direct print.

TEM: × 48 000.

Figs 36 and 37: *Scaphites rosenkrantzi* n. sp. (364-67) (Ammonoidea, *Scaphitaceae*) Upper Cretaceous, Senonian/Campinian, Scaphites naesen, The Agatdalen Nûgssuaq, West Greenland. Specimen given by Prof. Tove Birkelund.

Mineral composition of the samples: Aragonite Calcite, Gypsum.

Fig. 37 shows preservation of a dense network of short, cylindrical trabeculae, delimiting a rounded fenestration. In Fig. 38, the smooth cylindrical trabeculae are sturdier, the fenestration more elongate. The appearance of the structures in this field resembles more or less that of the nautiloid pattern.

Shadow-cast with platinum. Direct print.

Figs. 36 and 37: TEM: × 48 000

Plate 11.

Fig. 38: *Discoscaphites* sp. (388) (Ammonoidea, *Scaphitaceae*) Cretaceous, Fox Hills S.S., North of Little Eagle, Carson County, South Dakota, U.S.A. specimen given by Dr. N. S. Sohl (U.S.G.S. 21740) and Dr. A. G. Cooper.

Mineral composition of the samples: Aragonite.

In this deeply altered structure of the EDTA-insoluble conchiolin fraction of the nacreous layer of the shell wall, networks of flattened trabeculae have been transformed into a puzzle of indentated nodules. Groups of small spheroidal particles are scattered on the surface of these nodules, especially at their periphery.

Shadow-cast with platinum. Direct print.

TEM: × 48 000.

Figs 39 and 40: *Asteroceras stellare* (J. Sowerby) (929-1) (Ammonoidea, *Psilocerataceae*) Jurassic, Lower Lias, Charmouth, Dorset, England. Specimen given by Mr. S. Barney-Hansford.

Mineral composition of the samples: Calcite.

EDTA-insoluble conchiolin fraction of the nacreous layer of the shell wall. Worm-like, cylindrical and varicose, short, single or twisted, segments of trabeculae seem to be associated with extremely thin tortuous fibrils (asterisk), about 14 Å in diameter (see the structures at higher magnification in Fig. 40 with a magnifying lens).

Shadow-cast with platinum. Direct print.

TEM: Fig. 39: × 48 000; Fig. 40: × 56 000.

Fig. 41: *Amaltheus spinatus* Bruguiere (408) (Ammonoidea, *Eoderocerataceae*) Lower Jurassic, Lias delta, Banz, Franconia, Germany. Specimen given by Prof. O. H. Schindewolf.

Mineral composition of the samples: Aragonite.

EDTA-insoluble conchiolin fraction of the nacreous layer of the shell wall. In these two fields, the features of the nautiloid pattern are recognizable, in spite of a slight flattening and broadening of the trabeculae.
Shadow-cast with palladium. Direct sprint.
TEM: × 48 000.

Plate 12.

Fig. 42: *Hildoceras sp.* (1008) (Ammonoidea, *Hildocerataceae*). Jurassic, Lower Lias, Lyme Regis, Dorset, England.
Mineral composition of the samples: Calcite.
The different types of alteration listed in the description are assembled in these EDTA-insoluble residues of the mural conchiolin: flattened, cylindrical trabeculae, nodules, spheroidal and vesicle-like (asterisk) particles.
Shadow-cast with platinum. Direct print.
TEM: × 72 000.

Fig. 43: *Stepheoceras* sp. (989 CPC) (Ammonoidea, *Stephanocerataceae*). Lower Jurassic, Lower Lias, Whitby, Yorkshire, England.
Mineral composition of the samples: Calcite.
Nacreous layer of the shell wall fixed and decalcified by the CPC method. Twisted, tortuous, angulate (asterisk: centre right) and irregularly cylindrical trabeculae form loose networks which do not differ in appearance from the EDTA-insoluble residues of conchiolin. (See other micrographs of the same material in Grégoire, Goffinet & Voss-Foucart, in the press).
Shadow-cast with platinum. Direct print.
TEM: × 48 000.

Plates 13 and 14.

Figs 44 to 54 show, in the EDTA-insoluble fraction of conchiolin of the nacreous layer of the shell wall, spheroidal, pebble-shaped and vesicle-like particles, found in shells of ammonoids and of fossil nautiloids of different ages and horizons. These structures are shown in Devonian (Fig. 53), Pennsylvanian (Figs 44, 45, 46) and Jurassic (Fig. 47) nautiloids, in Paleozoic (Figs 48 and 49) and in Jurassic ammonoids (Fig. 51).
Spheroids and vesicles have been reproduced in conchiolin of modern, pyrolyzed mother-of-pearl (Figs. 50, 52 and 54.)

Fig. 44: *Michelinoceras directum* Unklesbay (1139-1) (Nautiloidea, Orthocerida). Pennsylvanian, Putnam Hill Limestone Member, Strip Mine, Canfield Township, Mahoning County, Ohio, U.S.A. Specimen given by Prof. J. L. Murphy.
Mineral composition of the samples: Aragonite.
In this sample the spheroidal particles are predominantly of small size (about 11-75 millimicrons).
Shadow-cast with platinum. Direct sprint.
TEM: ×48 000.

Fig. 45: *Solenochilus* sp. (1131-2) (Nautiloidea, Nautilida). Pennsylvanian, Brush Creek Limestone Member, Strip Mine Central Section 23, Madison Township, Columbiana County, Ohio, U.S.A. Specimen given by Prof. J. L. Murphy.

Mineral composition of the samples: Aragonite.

In this specimen the alterations in conchiolin did not differ in the brown glassy (this Fig.) and in the chalky portions of the shell wall (not shown). In this field, vesicle-like (160–170 millimicrons) structures are predominant over the spheroidal particles like those shown in Fig. 44.

Shadow-cast with platinum. Direct print.

TEM: × 66 000.

Fig. 46: *Liroceras liratum* (Girty) (1137) (Nautiloidea, Nautilida) Pennsylvanian, Kendrick Shale, Cow Creek, Floyd County, Kentucky, U.S.A. Specimen given by Prof. J. L. Murphy.

Mineral composition of the samples: Aragonite.

The predominant ultrastructures of conchiolin in this specimen were very small spheroidal particles of uniform size (about 14 millimicrons in diameter). Bottom centre and top left: intercrystalline cords.

Shadow-cast with platinum. Direct print.

TEM: × 36 000.

Fig. 47: *Cenoceras* sp. (949-2) (Nautiloidea, Nautilida). Lower Jurassic, Lower Lias, Region of Lyme Regis Charmouth, Dorset, England. Specimen given by Mr. S. Barney-Hansford.

Honey-brown nacreous layer of the shell wall (Calcite). Spheroidal particles of different sizes (23-114 millimicrons). The physical alterations in conchiolin did not distinctly differ in this sample, in which inversion of aragonite to calcite has taken place, and in still iridescent, aragonitic parts of the nacreous layer, in another region of the same shell.

Shadow-cast with platinum. Direct print.

TEM: 48 000.

Fig. 48: *Texoceras texanum* (Girty) 1908 (858) (Ammonoidea, Goniatitina) Late Lower Permian. Bone Spring Limestone, Culberson County, Texas, U.S.A. Specimen given by Prof. William M. Furnish and Prof. Brian F. Glenister.

Mineral composition of the samples: Calcite.

Spheroidal (20-190 millimicrons in diameter) and a few vesicle-like particles.

Shadow-cast with platinum. Direct print.

TEM: × 48 000.

Fig. 49: *Paraceltites elegans* Girty (856) (Ammonoidea, *Otocerataceae*). Late Lower Permian. Bone Spring Limestone, Culberson County, Texas, U.S.A. Specimen given by Prof. William M. Furnish and Prof. Brian F. Glenister.

Mineral composition of the samples: Calcite.

Spheroidal particles (33–130 millimicrons). A large, dark vesicle (380 millimicrons) is visible in the central part of the figure and two other vesicles at the bottom centre.

Shadowcast with platinum. Direct print.

TEM: × 42 000.

Fig. 50: *Nautilus pompilius* Linné. (1093-1 DB) Modern. Fragment of nacreous layer of the shell wall heated with sea water at 300 °C for 5 hours under a pressure of 500 bars per square cm.

Mineral composition of the samples: Calcite.

EDTA-insoluble conchiolin fraction. In this field, the thermally altered conchiolin consists of clustered spheroidal particles (26–160 millimicrons), identidal to those found in fossil conchiolin (see Fig. 49).
Shadow-cast with platinum. Reversed print.
TEM: 48 000.

Fig. 51: *Hildoceras* sp. (1008-5) (Ammonoidea, *Hildocerataceae*). Lower Jurassic, Lower Lias, Lyme Regis, Dorset, England.
Mineral composition of the samples: Calcite.
Clustered vesicle-like particles (about 70–240 millimicrons in diameter) are shown in different phases of disintegration (asterisk).
Shadow-cast with platinum. Direct print.
TEM: × 48 000.

Fig. 52: *Endolobus clorensis* Collinson (969-71) (Giant Nautiloidea, Nautilida). Carboniferous, Arkansas, U.S.A. (see also Figs 20 and 21). The largest vesicle-like particle in this cluster reaches 460 millimicrons in diameter (top left).
Shadow-cast with platinum. Direct print.
TEM: × 48 000.

Figs 53 and 54: *Nautilus pompilius* Linné (1092 DB) Modern. Fragment of nacreous layer of the shell wall heated with sea water at 300 °C for 5 hours under a pressure of 476 bars per square cm.
Mineral composition of the samples: Calcite.
Figs 53 and 54 show spheroidal, ovoid and vesicle-like structures (till 700 millimicrons and more in diameter), identical to those seen in Figs 45, 51, 52 and in altered conchiolin of ammonoids and fossil nautiloids from different ages and horizons, in which these structures can reach considerable sizes (see Fig. 24 and Grégoire, 1966a; Voss-Foucart and Grégoire, 1971).
Shadow-cast with platinum. Reversed prints.
TEM: × 48 000.

Plate 15.

Fig. 55: (see Note 8) *Nautilus pompilius* Linné (978) Modern. Nacreous layer of the shell wall heated with sea water in sealed tubes at 800 °C for 5 hours (Grégoire, 1972b).
Mineral composition of the samples: Calcite.
An EDTA-insoluble conchiolin residue appears in this field in the form of three small and a large, hexagonal structures with straightlined boundaries. Networks of flattened trabeculae and nodules are contained in the large hexagon. Alignements of nodules superimposed on to the networks, are assembled in straight rows parallel to the sides of the large hexagon. This complex organic structure (crystal "ghost"; Grégoire, 1968; see also Figs 124 and 125 in Grégoire, 1972b) seems to have been produced by squeezing and moulding of interlamellar conchiolin between euhedral facets of thermally expanding polyhedral crystals during inversion of aragonite to calcite.
Shadow-cast with platinum. Direct print.
TEM: × 57 000.

Fig. 56: *Nautilus pompilius* Linne (813) Modern. Nacreous layer of the shell wall heated in quartz tubes sealed under vacuum at 600 °C for 5 minutes (Preheating time, 20 °C to 600 °C: 15 minutes) (Grégoire, 1968).

Mineral composition of the samples: Calcite.

EDTA-insoluble conchiolin fraction. These alterations in the form of rings and hairpins (asterisk) resemble the thermal changes in the form of toroids and hairpins observed in polypeptides by Blais and Geil (1968). Compare Fig. 56 with Fig. 3 F of Blais and Geil: polybenzyl-glutamate (see section 4-4-4). In the conchiolin of modern nacreous layer pyrolyzed at high temperatures, pads, which are local thickenings, are produced by retraction of substance around the holes in the perforated membranes of coalescence of the trabeculae (see Grégoire, 1968, Figs 68 and 69; 1972b, Fig. 106). The structures shown in Fig. 56 might be pads detached from the perforated membranes.

Shadow-cast with palladium. Direct print.
TEM: × 54 000.

Fig. 57: (see Note 5) *Nautilus pompilius* Linné (1100) Modern. Nacreous layer of the shell wall heated with distilled water in sealed tubes at 160 °C for 53 days.

Mineral composition of the samples: Aragonite.

These feltworks of fibrils, associated with spheroidal particles, might be residues of conchiolin left after an important extraction of soluble substances during a protracted exposure to water (see Grégoire 1972b and Voss-Foucart and Grégoire, 1975).

Shadow-cast with platinum. Direct print.
TEM: × 48 000.

Fig. 58: *Dolorthoceras sociale* (Hall) (677-76) Nautiloidea, Orthocerida). Upper Ordovician, Lower Maquoketa Formation, Graf, Dubuque County, Iowa, U.S.A. Specimens given by Prof. W. M. Furnish and Prof. B. F. Glenister.

The EDTA-insoluble fraction of the organic substance which separates or surrounds the diverging elongate crystals of the spherulites (see Plate 2) appears in the form of soft fragments of ribbons.

Shadow-cast with platinum. Direct print.
TEM: × 48 000.

Fig. 59: (see Note 3) *Nautilus pompilius* Linné. Modern (DB 4-740). Nacreous layer of the shell wall compressed at room temperature (20 °C) for 5 hours at 50 000 kgs per square cm.

Mineral composition of the samples: Aragonite.

Direct replica (Carbon-Platinum) of a surface of cleavage and of fracture.

In Fig. 59, the surface of a polygonal crystal shows a double orientation at an angle of 60°–70°, of straight rows of tubercular elevations (asterisks).

TEM: × 30 000

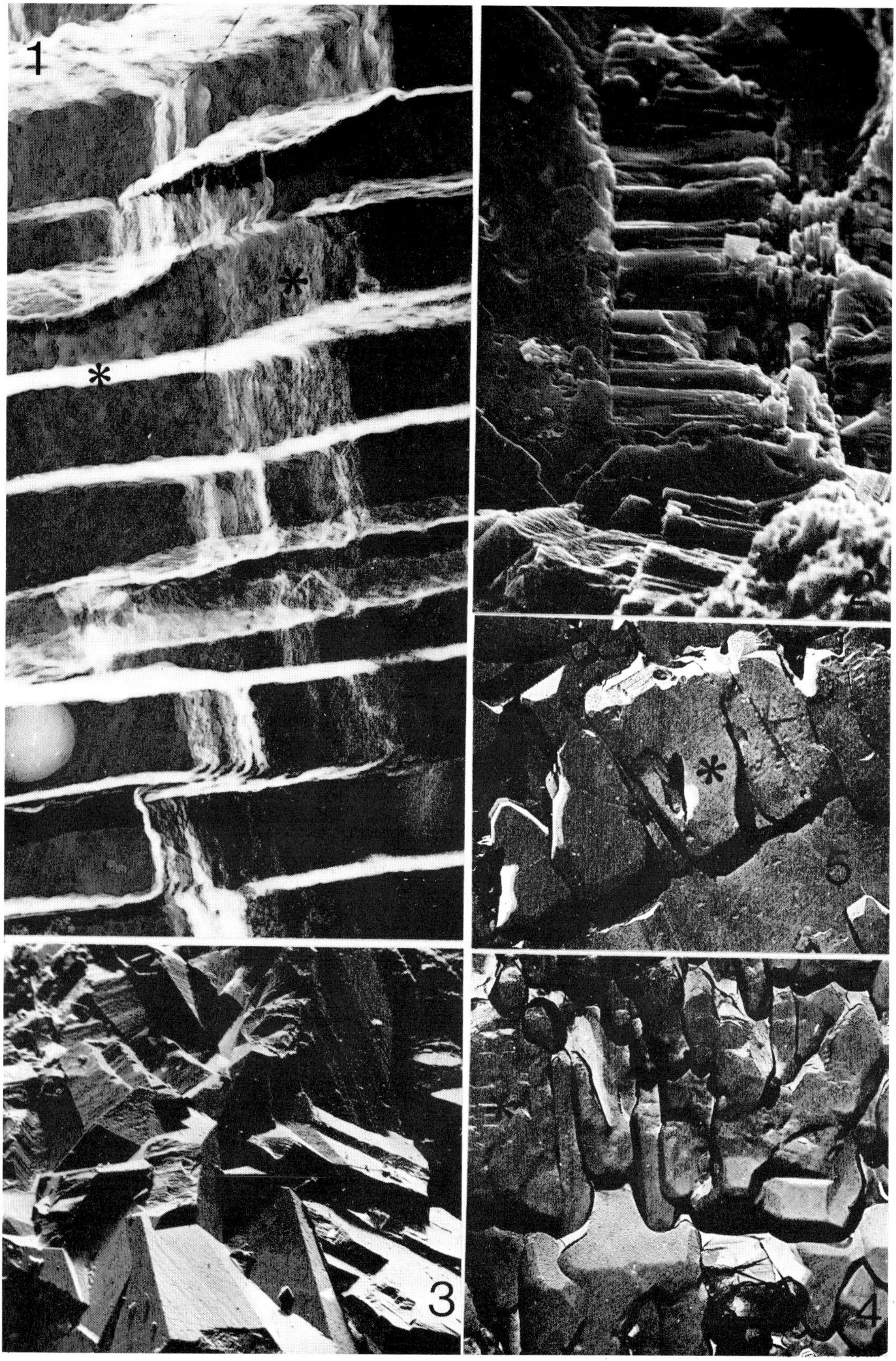
1
2
5
3
4

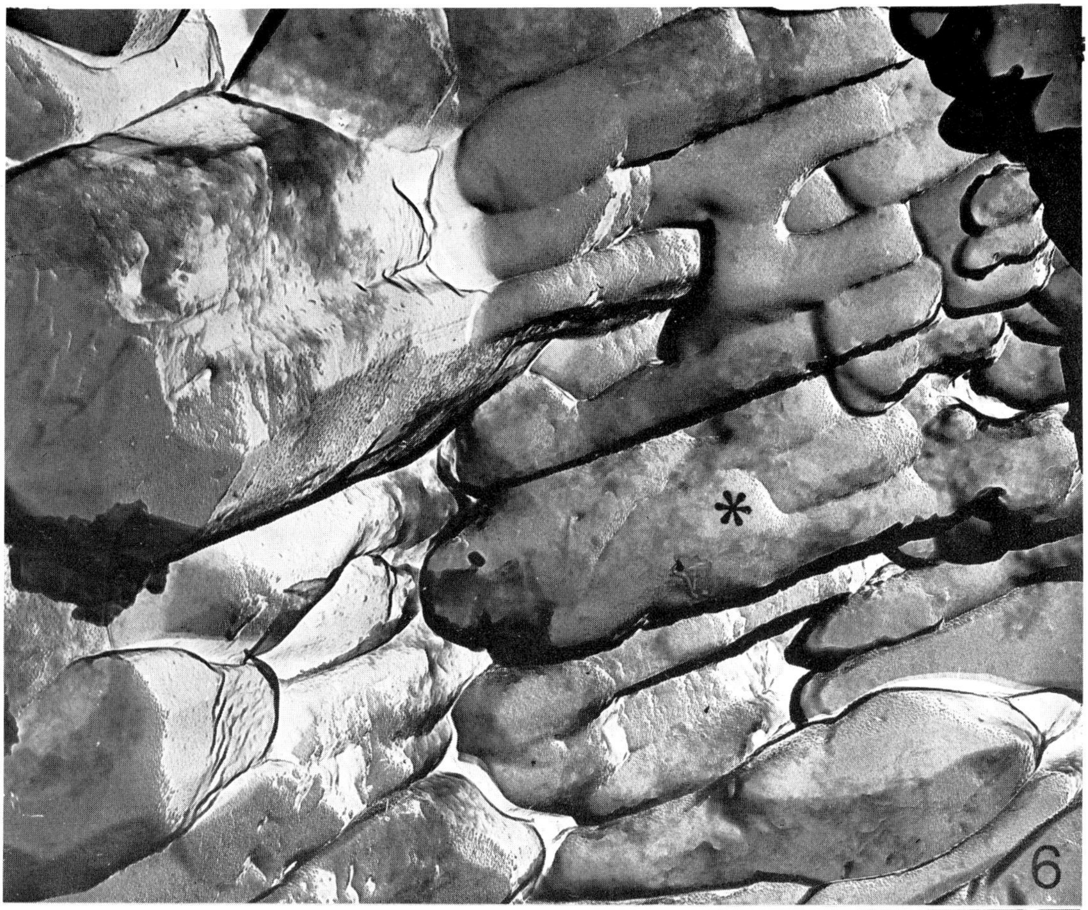
*
6

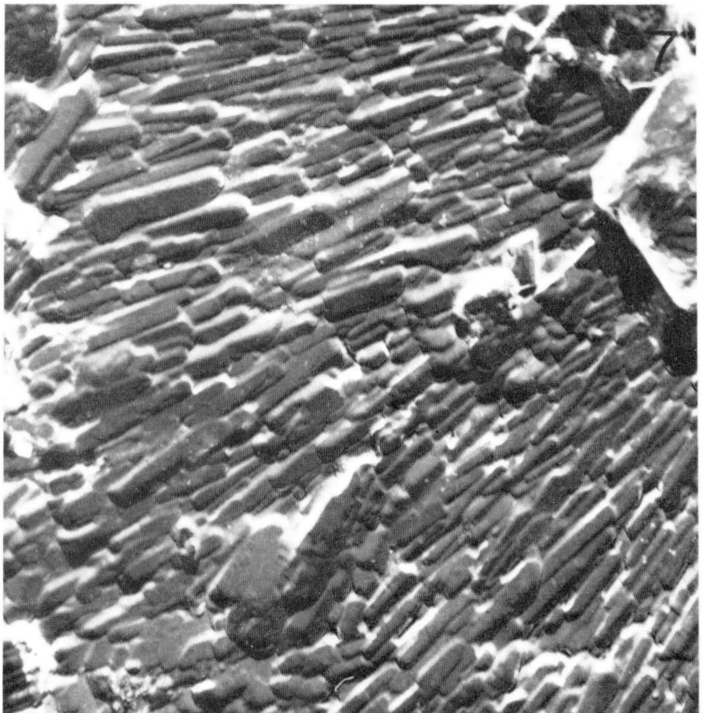
7

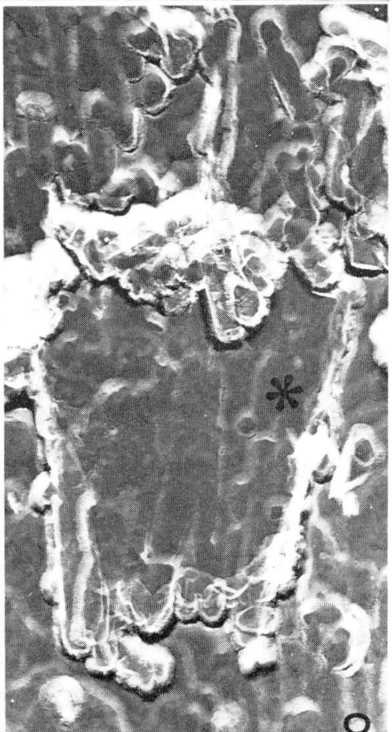
*
8

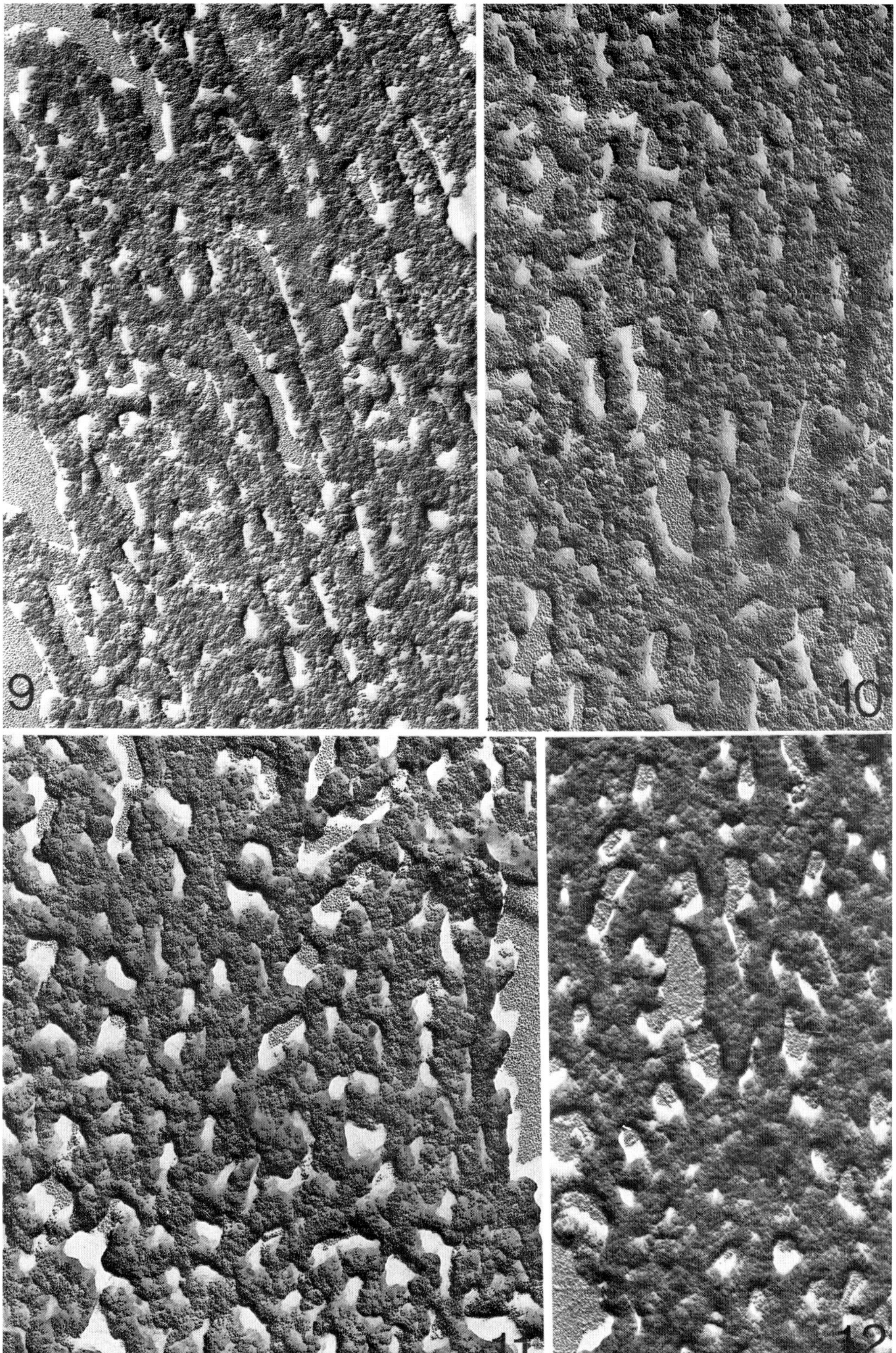
9
10
11
12

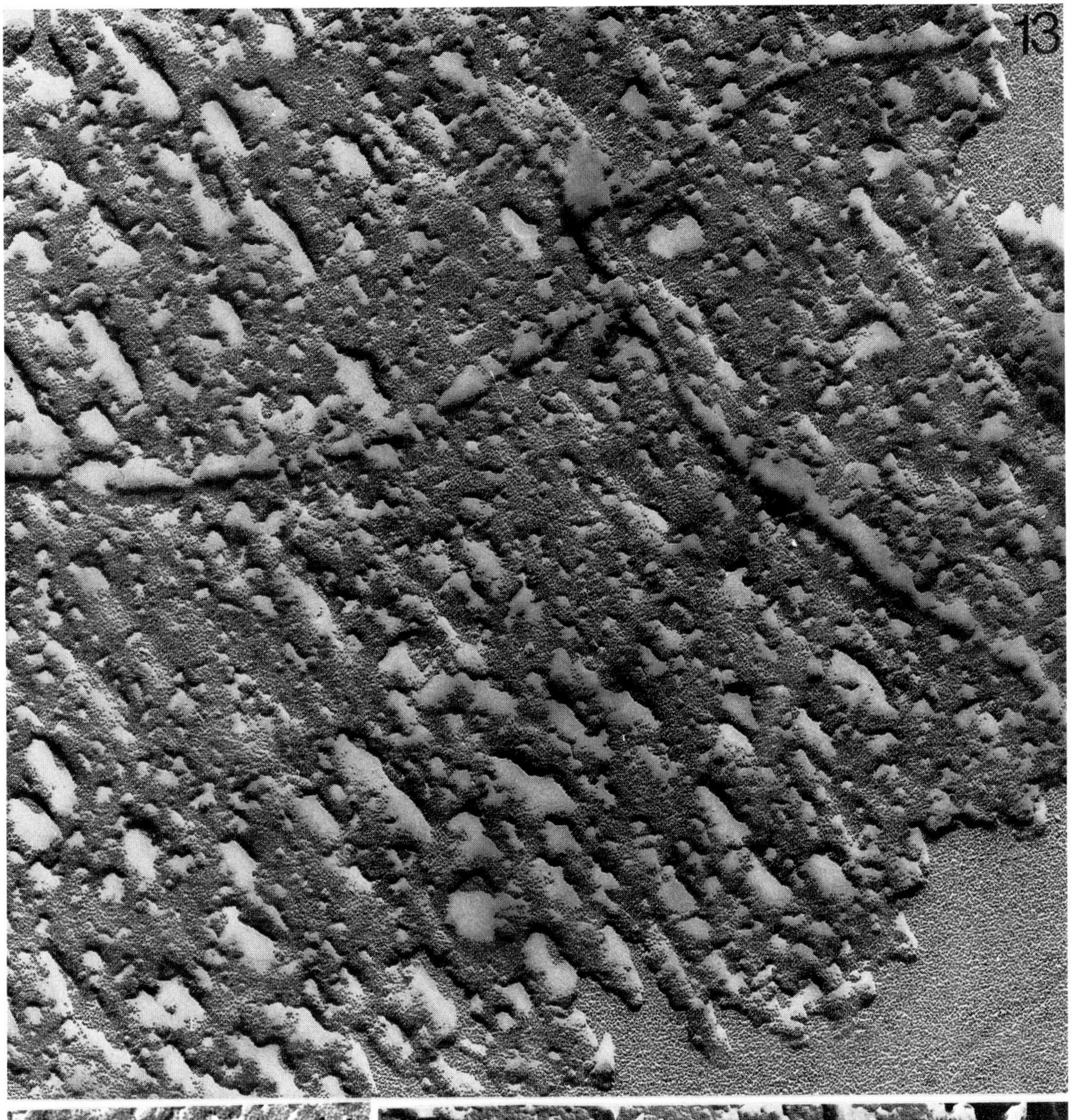
13

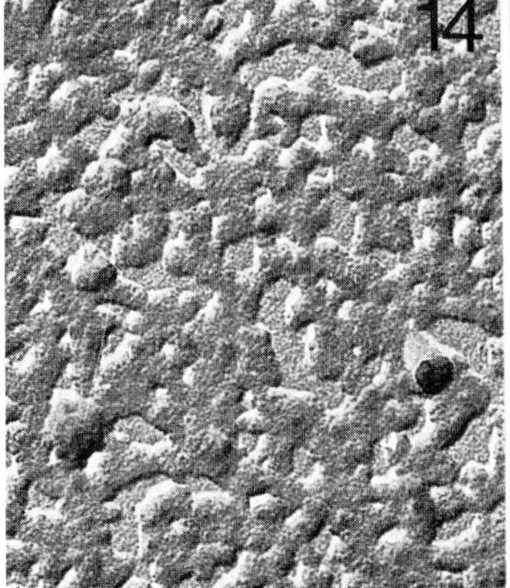
14

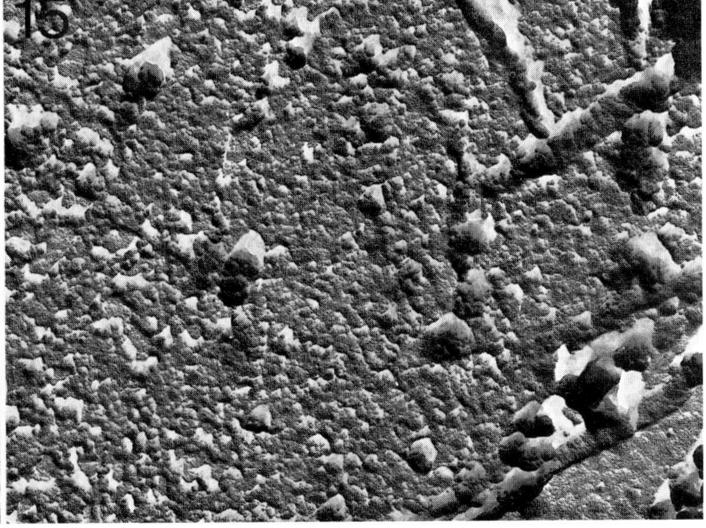
15

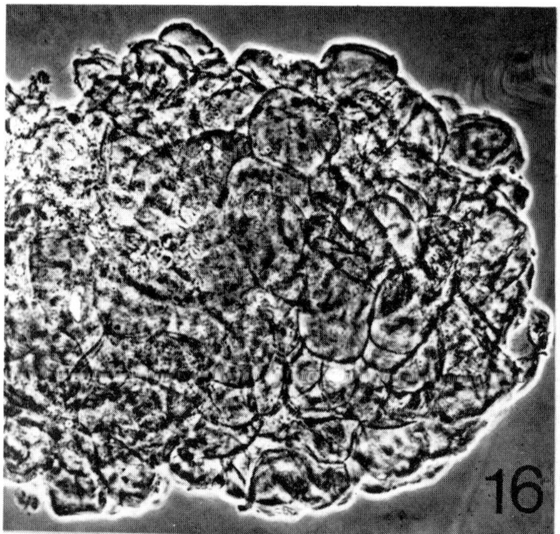

16

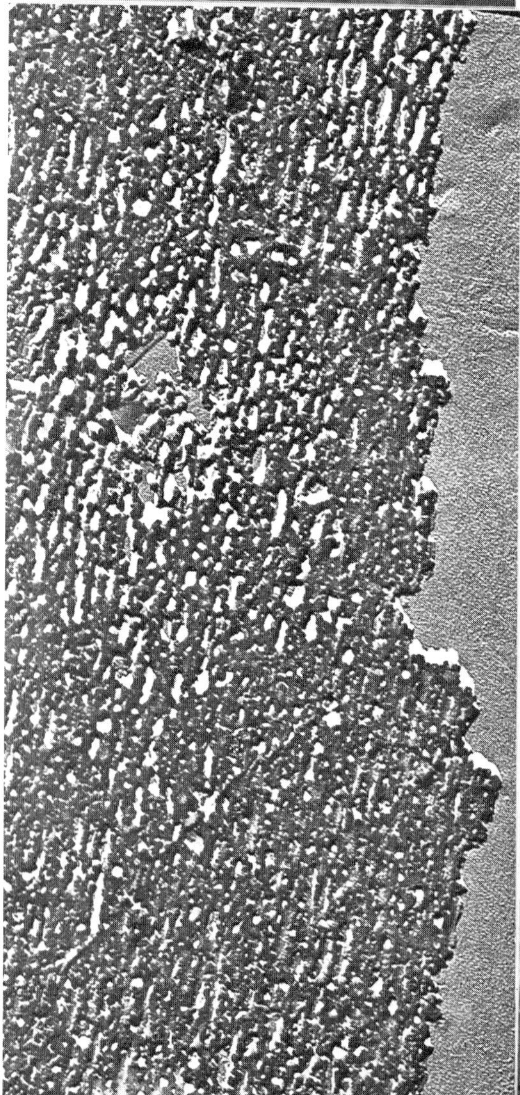

17

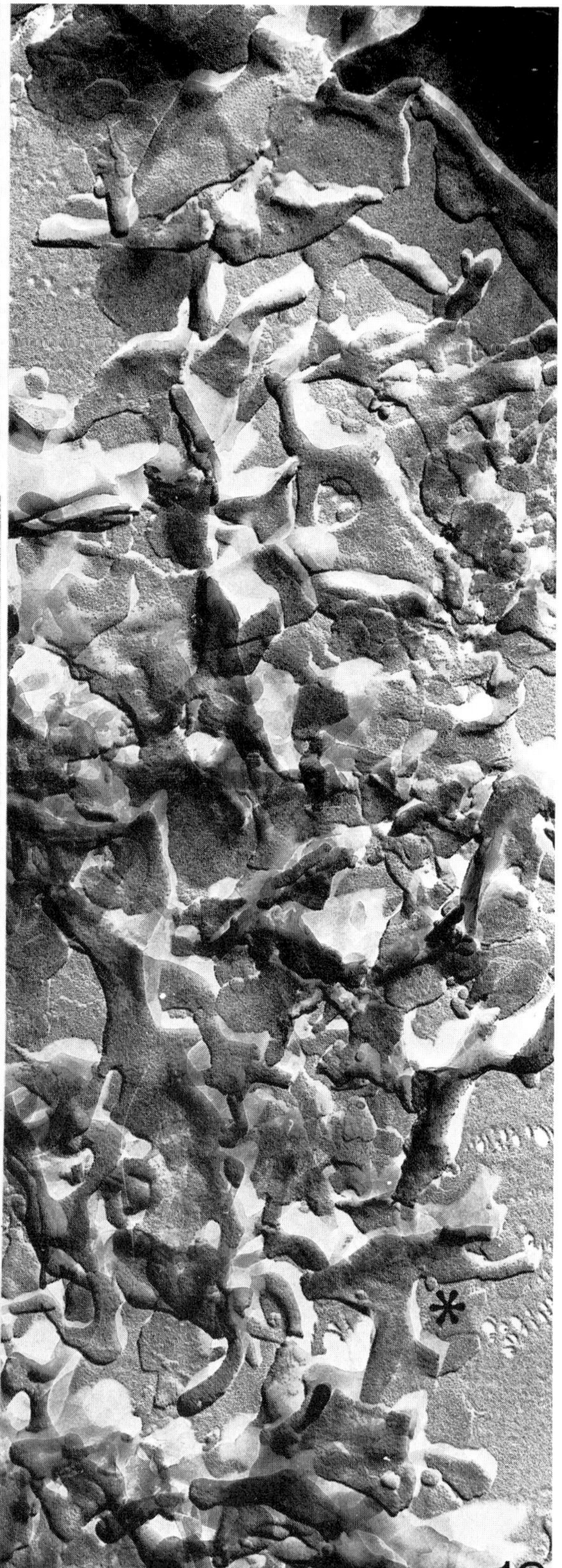

18

19

20

21

22
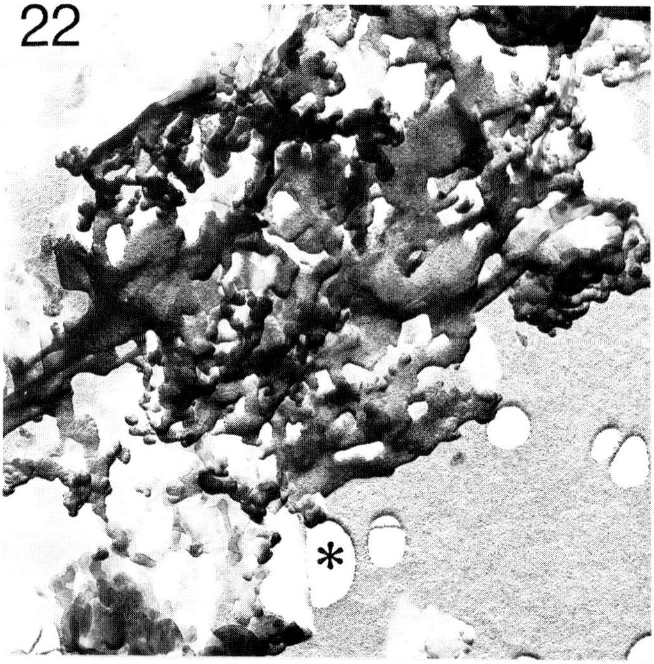

23
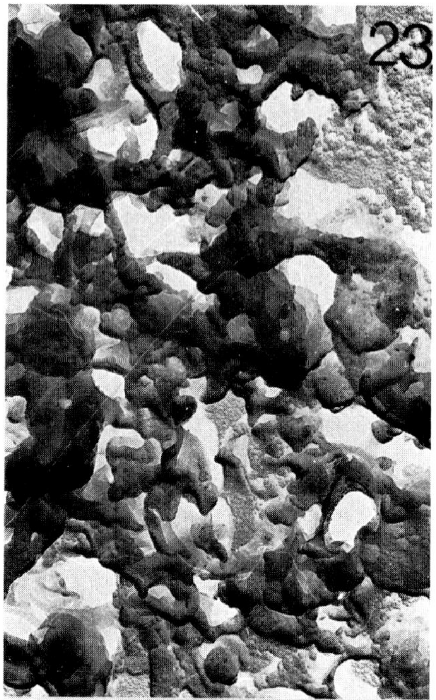

24
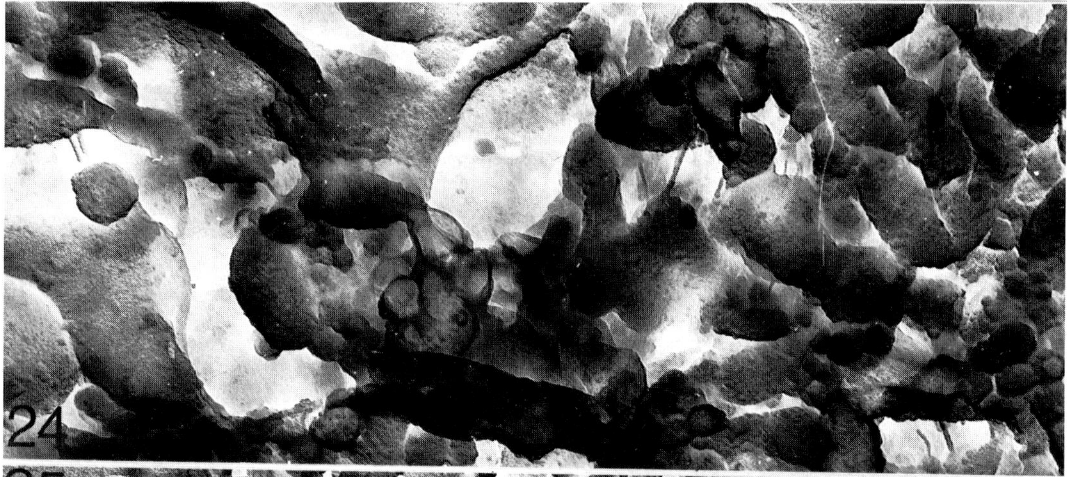

25
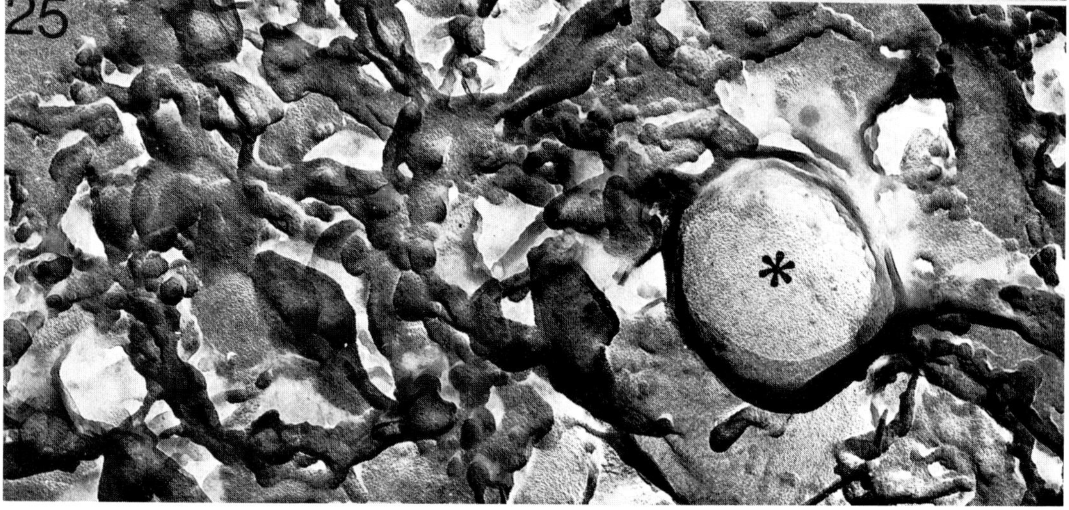

26

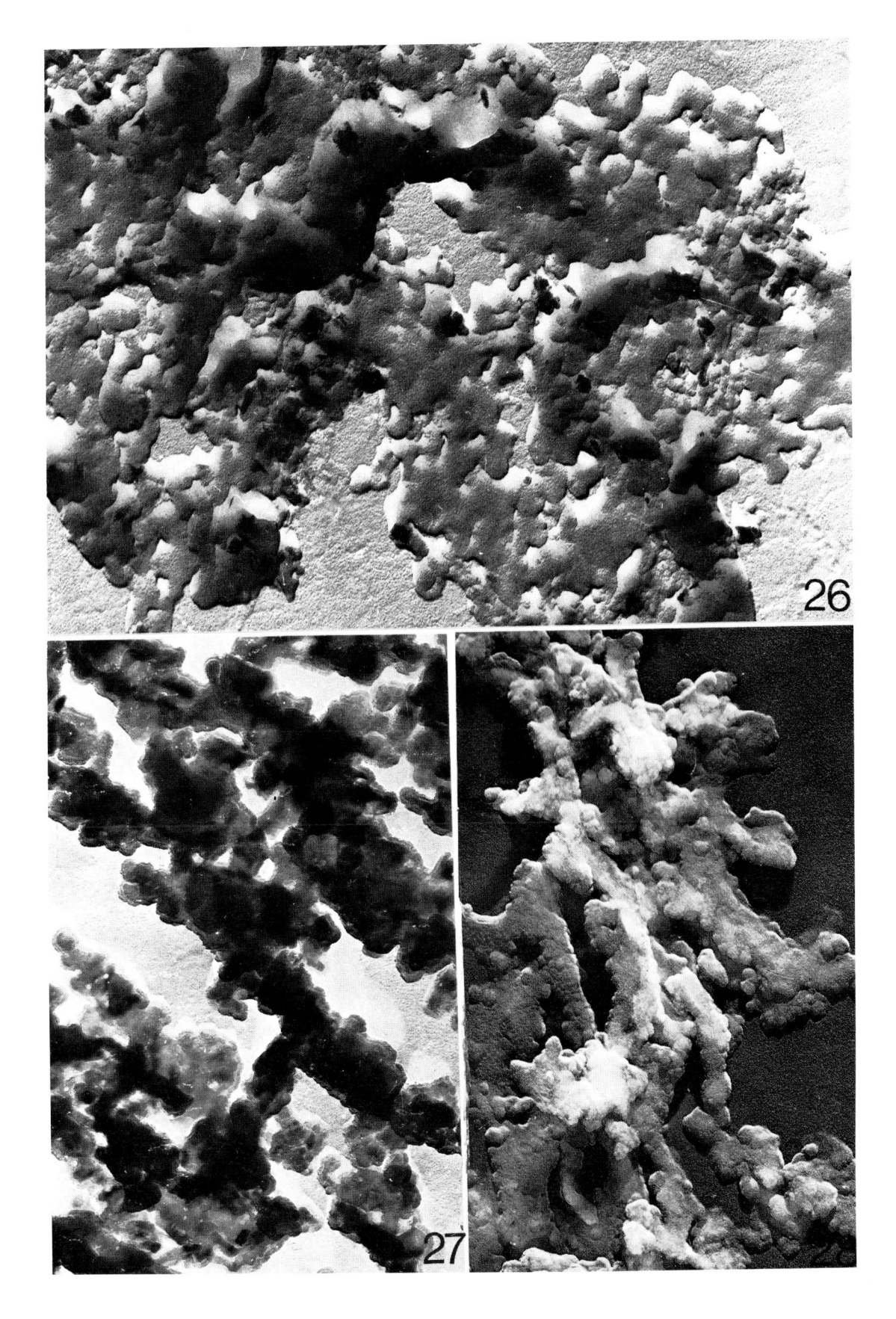

27

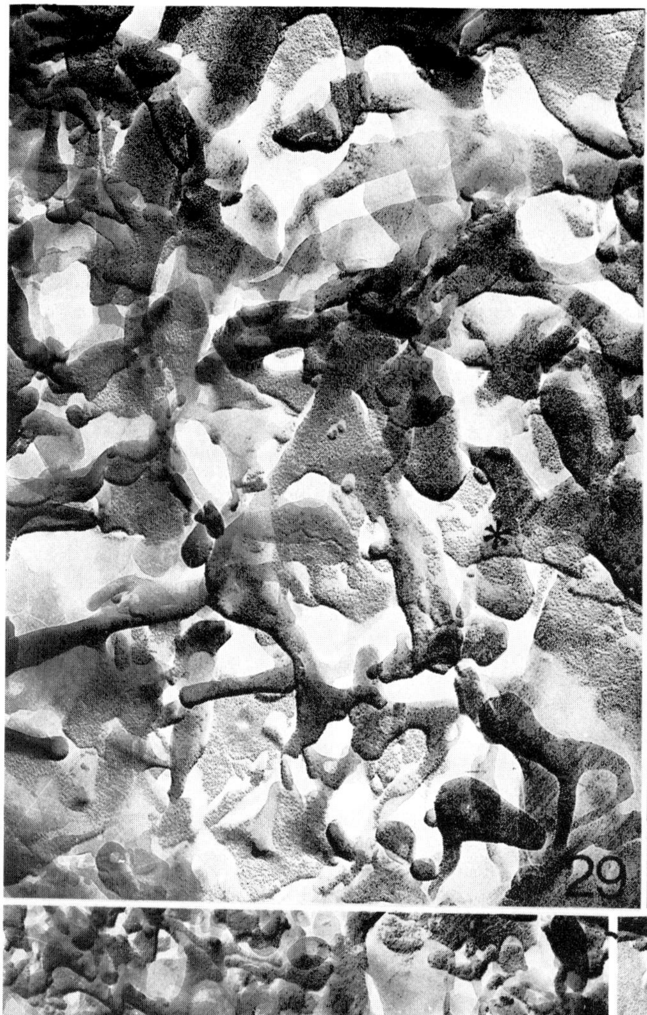
29
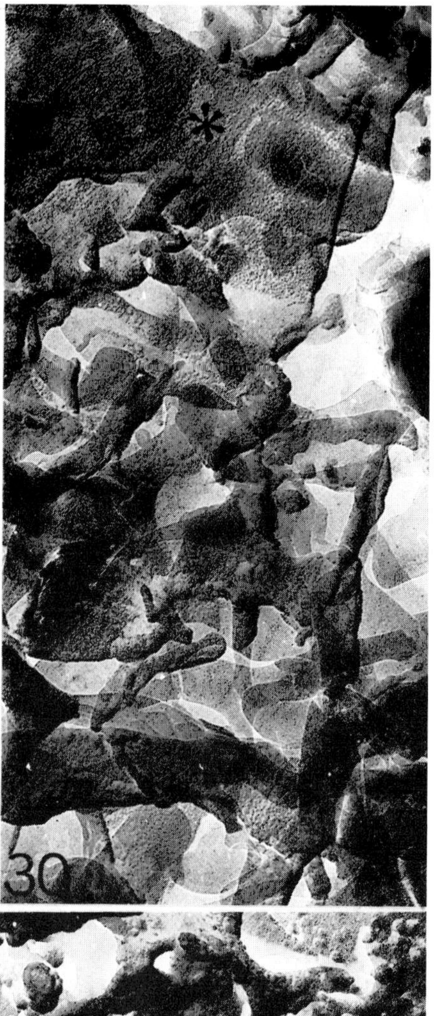
30
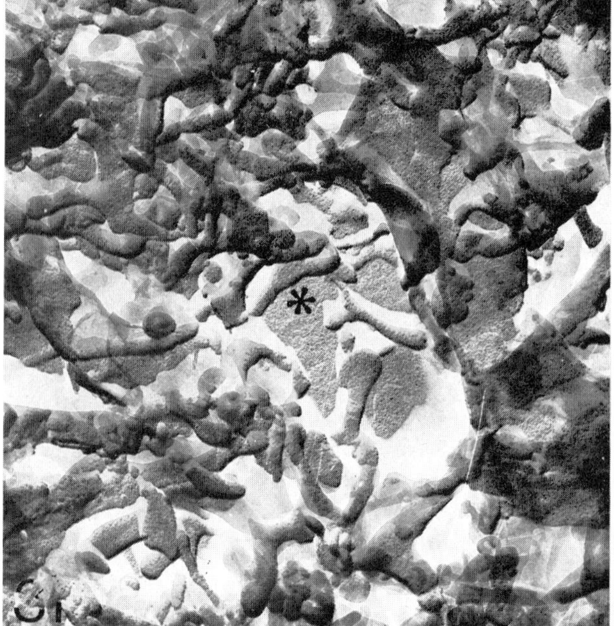
31
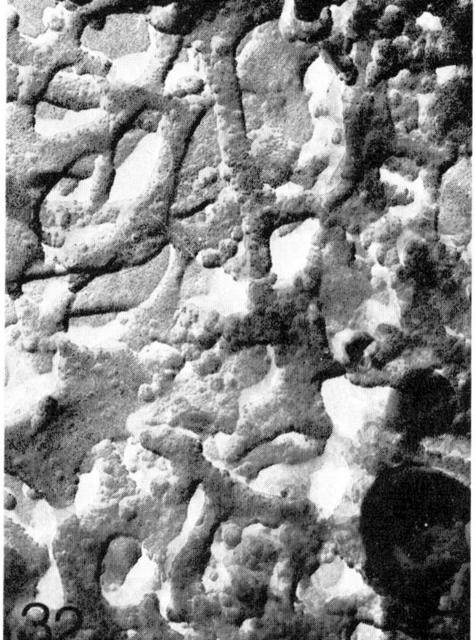
32

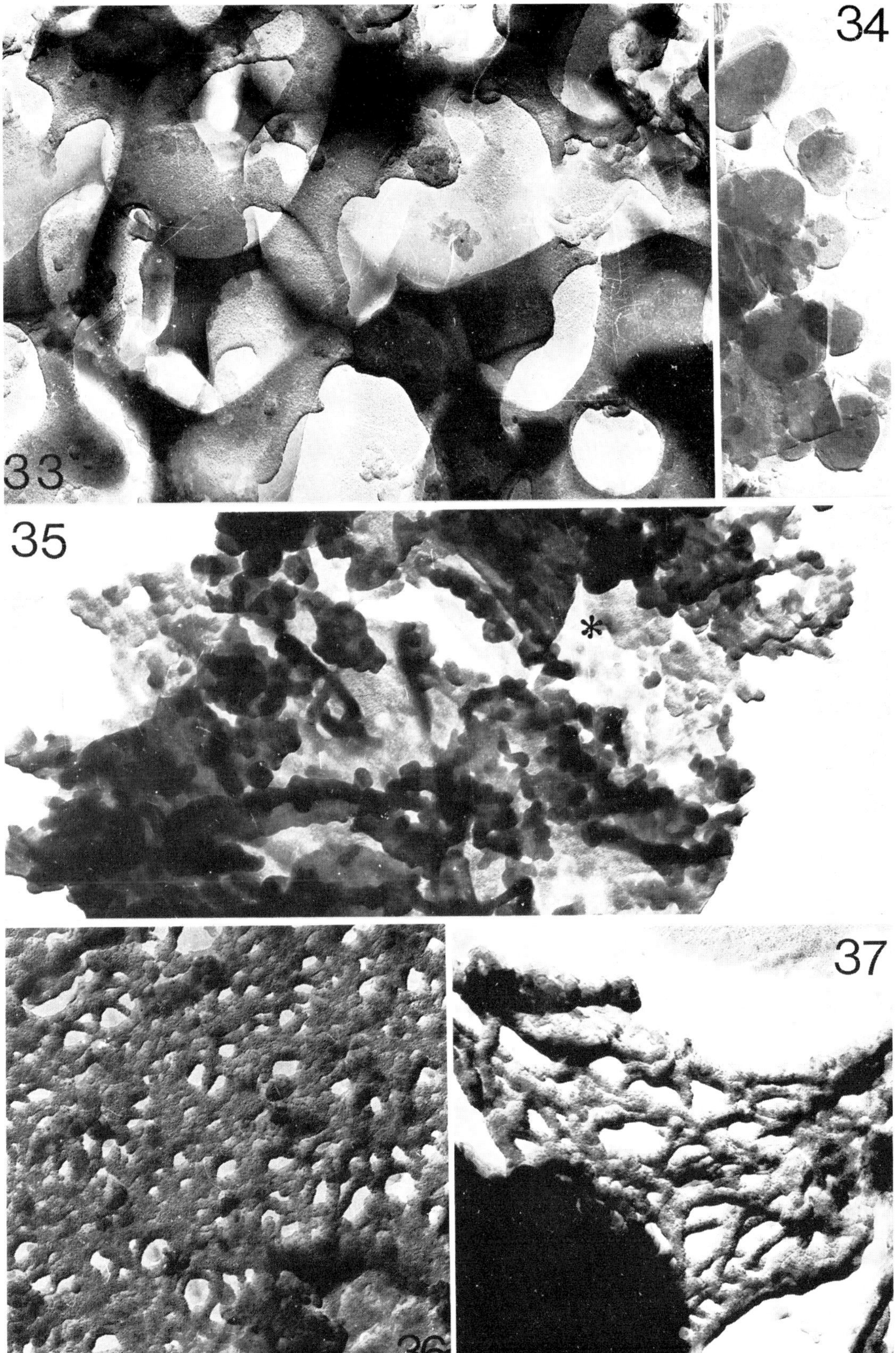
34
33
35
*
37
36

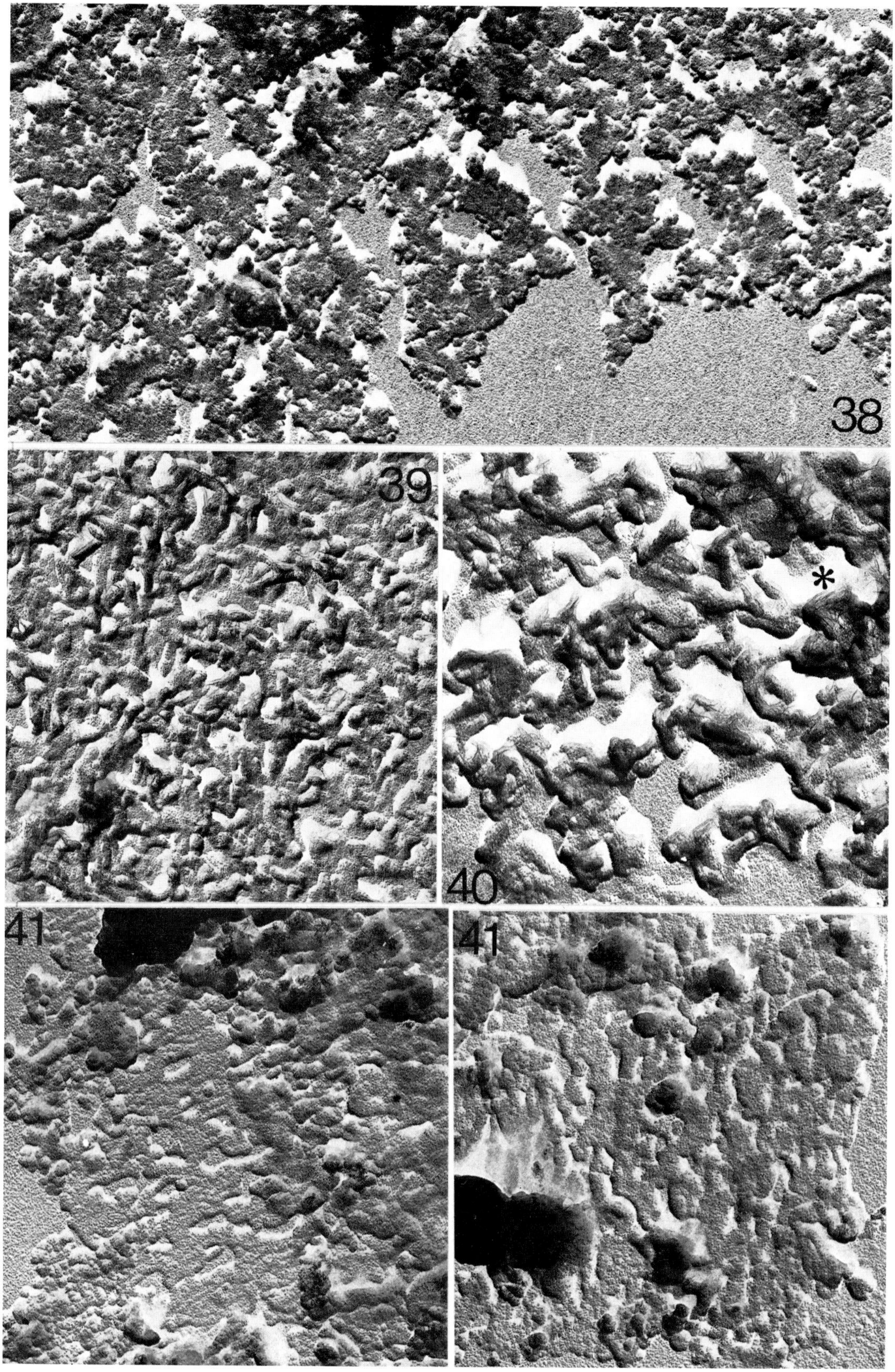
38
39
40
41
41

42

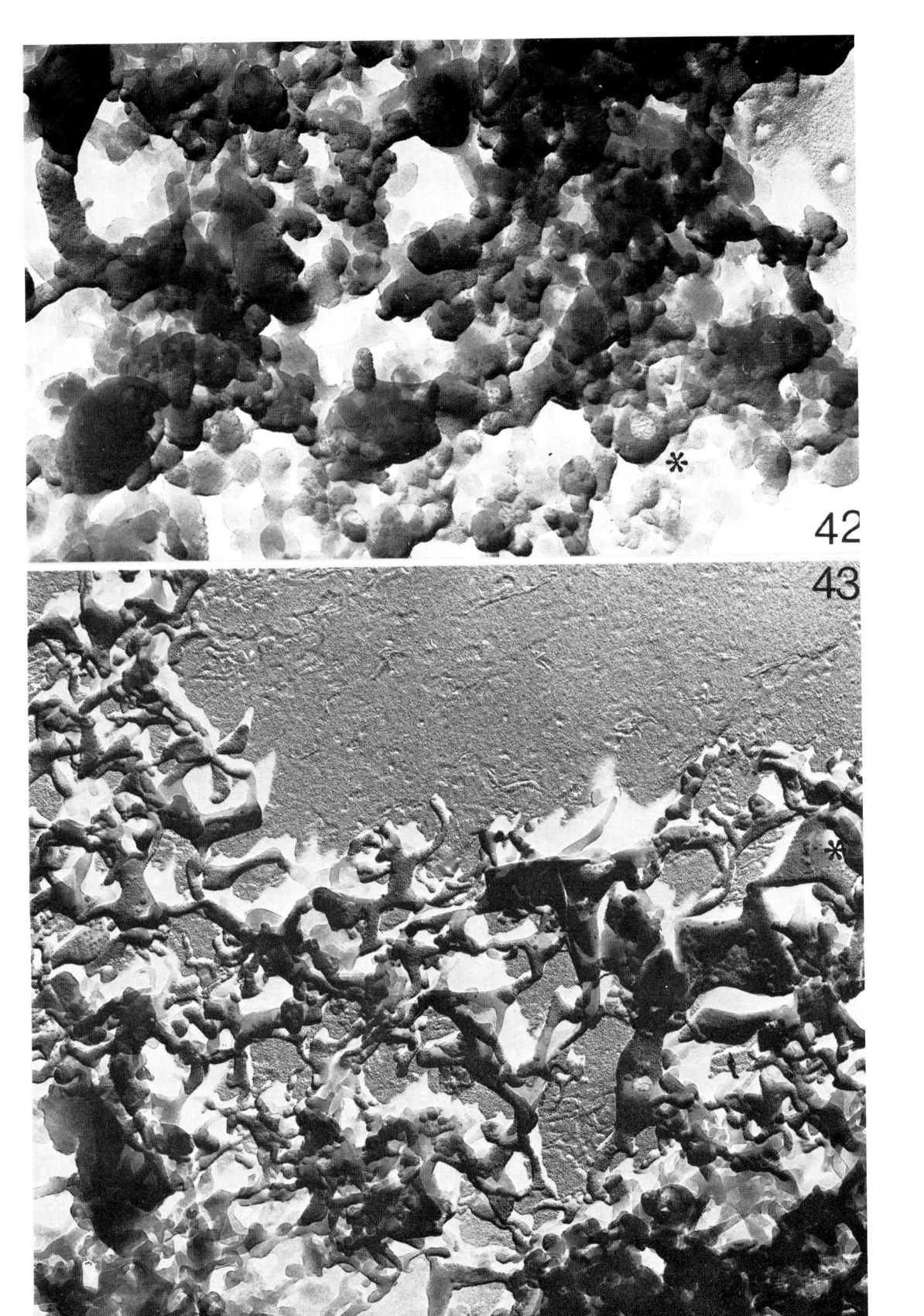

43

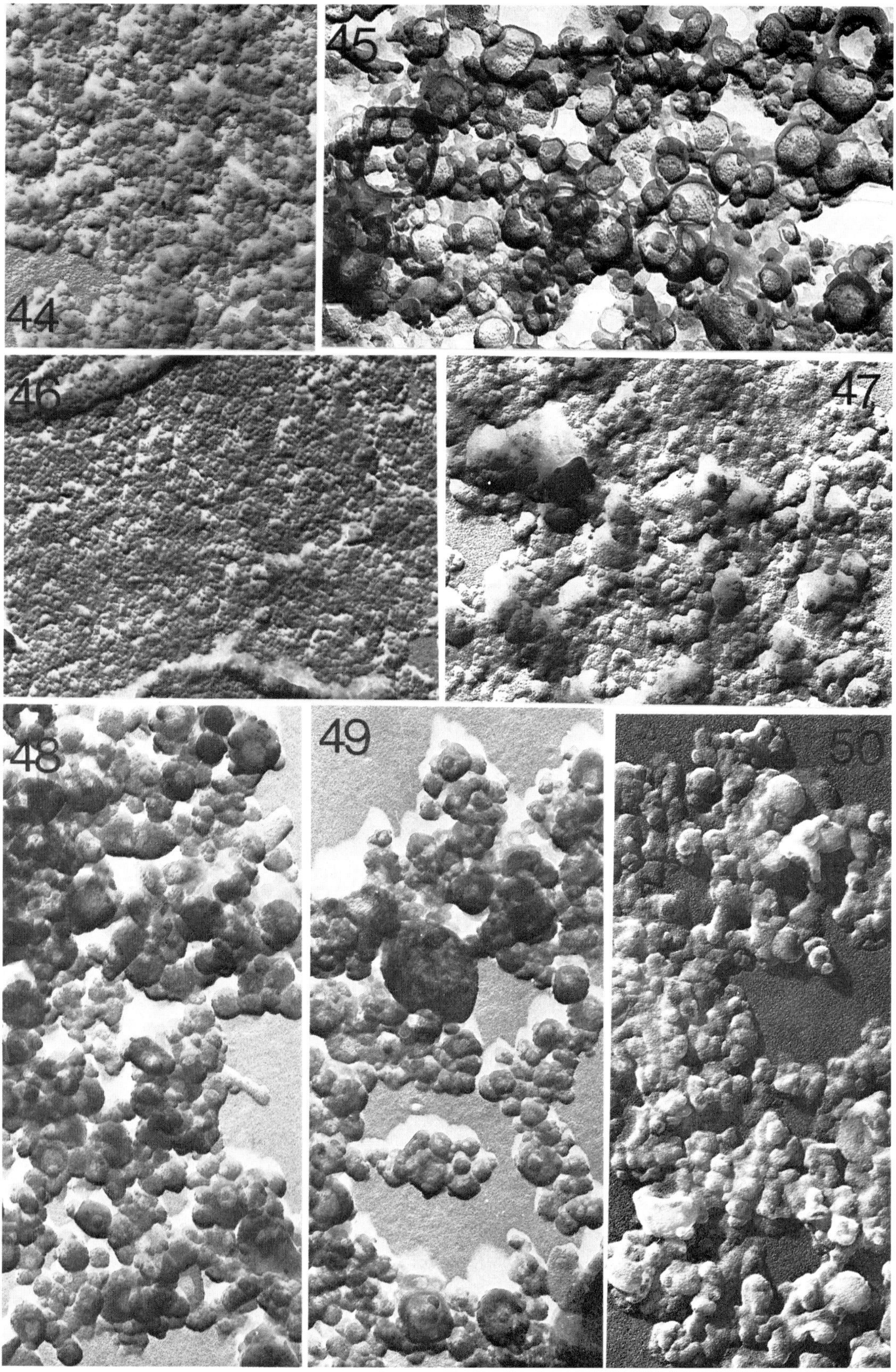
44
45
46
47
48
49
50

51

52

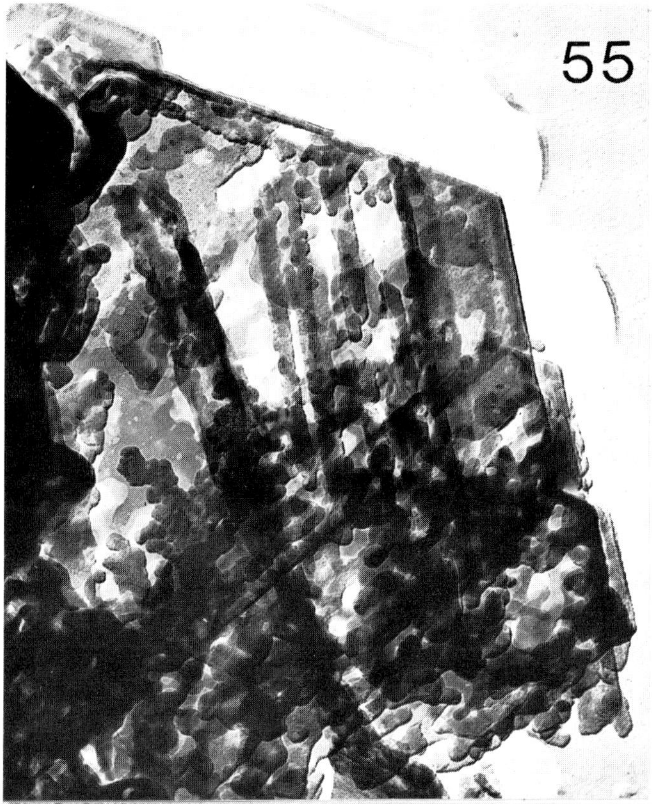
55

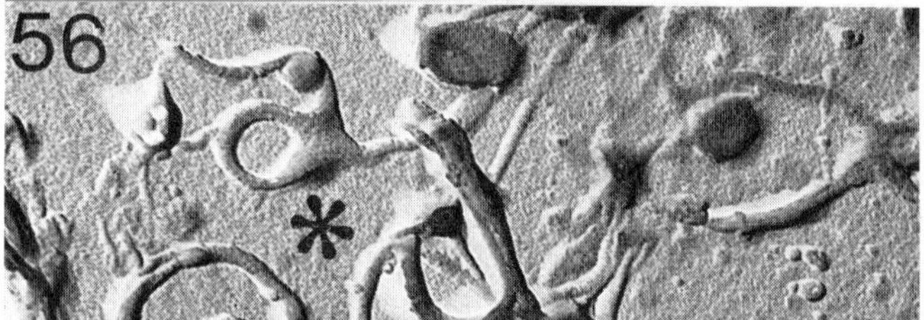
56

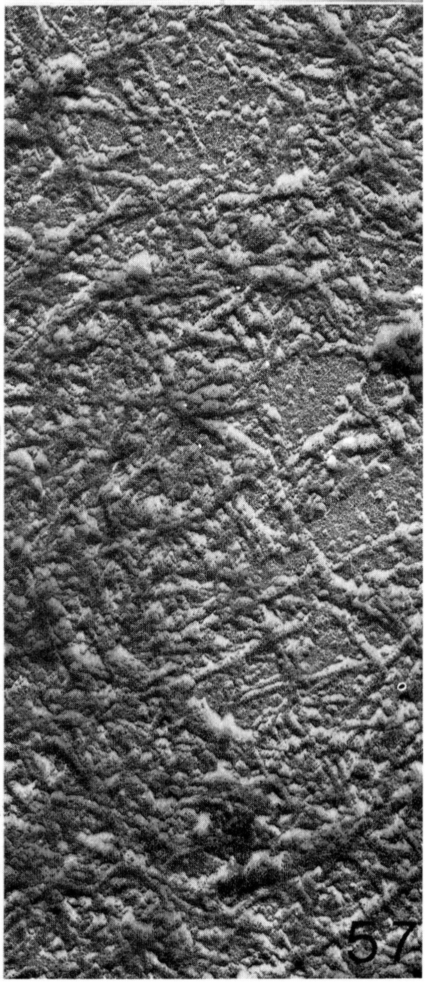
57

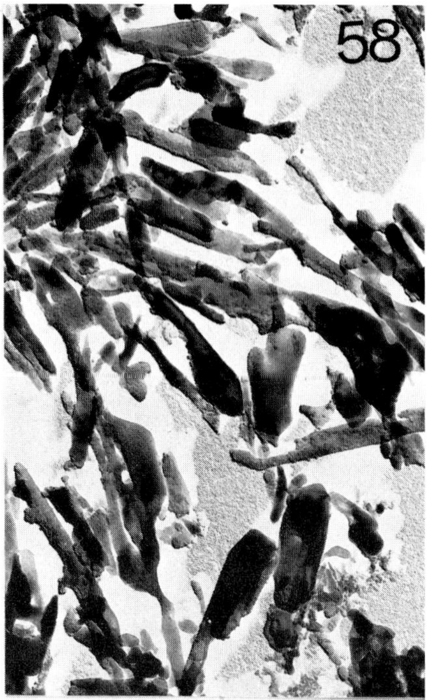
58

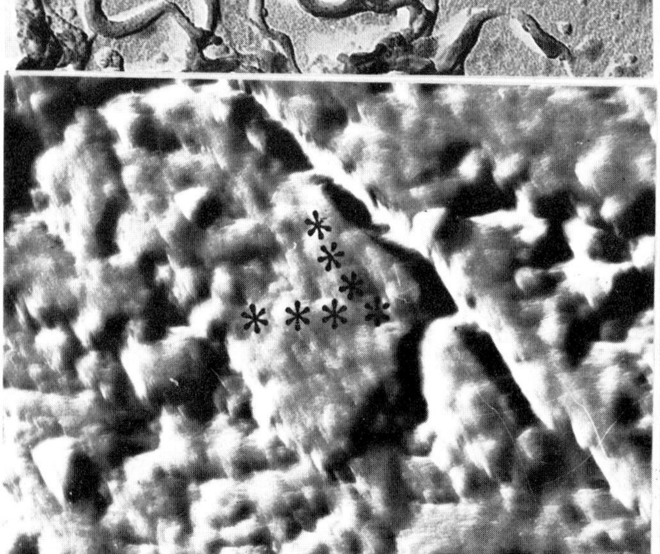
59

10. TABLES

Grades in intensity of the biuret reaction in the conchiolin shreds (pink violet flakes) (Fig. 16) are indicated by one, two or three asterisks.

Empty compartments (4 th column) mean that no type of structural alteration was predominant in the samples.

Predominance of one or more types of alteration in other compartments is an average estimate on the basis of examination of a great number of micrographs.

The following abbreviations have been used (fourth column):

n p (nautiloid pattern) (Type 1)

am p (?): possible ammonoid pattern.

cyl (indrical) varic (ose) trabeculae (Type 2)

Sph (eroidal) particles (Type 3), frequently assembled in clusters with a botryoidal appearance.

ves(icle – like) particles (Type 4).

fl(attened), broa(dened) angul(ate) trabeculae (Type 5)

mb (membranes) and their fragments sl(abs), nod(ules) (Type 6):

rib (ribbons).

Former papers containing illustrations of the material: G(régoire); GT (and Teichert); GVF (G. and Voss-Voucart); VFG (Voss-Foucart and G.); GL (G. and Lorent); GGL (Grandjean, G. and Lutts).

"Chalky": differs from nacreous and from crystalline. A chalky shell wall appears like chalk; it is very hard or soft, porous, powdery. It can be iridescent or lustreless. Where it is caused by weathering, a chalky material can conceal underlying iridescent nacreous substance.

Table 1: NAUTILOIDS (1)

MATERIAL	STRATIGRAPHY Age, Horizon, Locality	MINERALOGY (Nacreous layer)	Biuret Reaction (❖) Predominant type of conchiolin alteration. References of former illustrations	Description of the samples
O. ENDOCERIDA				
Endoceras vaginatum Schl. sp. (793)	Lower Silurian B3 (*Ordovician*) Reval, Estland	Calcite	❖❖ mb sl	Brown Crystalline
Endoceras sp. (961)	*Lower Ordovician* Arenig (Grauer Orthoceren-Kalk) Yxhult, Schonen, Southern Sweden	Calcite	❖ mb sl	Brown Crystalline
Endoceras n. sp. (454)	*Upper Ordovician* Black River Formation Montréal, Québec, Canada	Calcite	❖ cyl varic sph ves	Black-brown Crystalline
O. ACTINOCERIDA				
Armenoceras barnesi Flower, n. Sp. (Flower 1957, Pl. 9, Fig. 7) (701)	*Upper Ordovician* Burnam Limestone Burnet County, Central Texas, U.S.A.	Calcite (ray 3.03 Å); Quartz & unidentified substances.	❖ cyl varic sl	Yellowish Crystalline

Orthonybyoceras duseri (705) Flower, 1957, 1964	*Upper Ordovician* Cincinnatian Waynesville beds, Cincinnati Region, Ohio, U.S.A.	Calcite	❊❊ G 1967	Grey Crystalline
Treptoceras sp. (*Orthonybyoceras* Shimizu & Obutu, 1935) (1050)	*Upper Ordovician* Cincinnatian	Calcite	❊❊ cyl varic sl sph	Dark-grey-brown Crystalline
Rayonnoceras sp. (1203) (Saunders, 1973)	*Lower Carboniferous* Mississippian Fayetteville Shale, Arkansas, U.S.A.	α-quartz	Fig. 18 varic angul	Black Crystalline

O. ORTHOCERIDA

Orthoceras alveolare Hauer (787)	*Triassic* Hallstadt, Austria	Calcite	❊ mb	Brown Crystalline
Michelinoceras sp. (Geological Survey of Western Australia F 5243) (1052)	*Upper Devonian* Fairfield Formation Oscar Hill, West Kimberley, Australia.	Calcite	angul mb	Yellow-orange Crystalline

Michelinoceras sp. (Geological Survey of Western Australia Nr 19951) (1064 to 1067)	*Upper Devonian* Virgin Hills Formation, east (19951) and west (21551) sides of McWhae Ridge, Bugle Gap Area, Kimberley Division, Western Australia.	Calcite	fl mb sl sph	Red-brown and white Crystalline with red coating
Michelinoceras sp. (Geological Survey of Western Australia Nr 21551) (1068, 1069)		Calcite	Textfig. 3 a b	Orange-red or light-pink-brown Crystalline
Michelinoceras directum Unklesbay (1139) (J. L. Murphy, 1970)	*Pennsylvanian* Putnam Hill Limestone Member, near Canfield Township, Mahoning County, Ohio, U.S.A.	Aragonite	Fig. 44 mb sph	Coal-black Crystalline
Kionoceras ungeri (Sturgeon & Miller) (1136) (J. L. Murphy, 1966)	*Pennsylvanian* Putnam Hill Limestone Member, West of Canfield Township, Mahoning County, Ohio, U.S.A.	Aragonite	Traces n p mb sph	Pink Chalky
Bitaunioceras sp. (857) (Flower, 1941)	latest *Lower Permian* Bone Spring Limestone Culberson county, Texas, U.S.A.	Calcite	❖	Dark-brown Crystalline

The conchiolin matrices in nacreous layers

Leurocycloceras bucheri (Flower) (698)	*Upper Ordovician* Laurel Limestone Westport, Indiana, U.S.A.	Calcite	❖❖	Yellow-brown Crystalline
Striacoceras typus (Saemann) (Flower, 1936, 1939) (703)	*Middle Devonian* Cherry Valley Limestone Stockbridge Falls, New York, U.S.A.	Calcite	❖❖ cyl varic mb sl ves sph G 1966a	Black Crystalline
Dawsonoceras annulatum Sow. (792)	*Upper Silurian* Gotland, Sweden	Calcite	❖ sl	Grey crystalline
Pseudorthoceras knoxense McChesney about 40 specimens (361, 390, 421, 530, 720 1202) (Unklesbay, 1962)	*Lower Middle Pennsylvanian* Buckhorn Asphalt Sulphur, Oklahoma, U.S.A.	Aragonite Aragonite & ray 3.03Å of Calcite Navajoite (530)	❖❖ n p cyl varic mb sph GT 1965; G 1966a, 1968; GL 1972	Dark-brown (asphalt) Nacreous iridescent crystalline
Mooreoceras normale cf. var. *uniconstrictum* Miller & Owen (1133) (J. L. Murphy, 1970)	*Pennsylvanian* Putnam Hill Limestone Member, Strip Mine, Canfield Township, Mahoning County, Ohio, U.S.A.	Aragonite	Sph ves	Pink Chalky

Taxon	Age / Locality	Minerals	Figures	Description
Loose, curved flakes from nacreous layers of large, orthoconic (or coiled) nautiloids. (385, 390, 361, 421, 530, 720: about 120 samples) Orthoconic brevicone (607)	*Lower Middle Pennsylvanian* Atoka Formation Buckhorn Asphalt, Sulphur, Oklahoma, U.S.A.	Aragonite Aragonite & Calcite (ray 3.03Å)	❖❖ Figs 13, 15 and 17 np cyl varic sph G 1959 b; GGL 1964; GT 1965 GVF 1970; VFG 1971 G 1966a, 1968, 1972	Dark-brown (asphalt) Translucid Nacreous and crystalline 607: brittle strongly iridescent with intense metallic hues (silvery, blue, violet flakes of cleavage)
Dolorthoceras sociale (Hall) (357, 360, 677) (Tasch, 1955)	*Upper Ordovician* Lower Maquoketa Formation Graf, Dubuque County, Iowa, U.S.A.	Navajoite (357) Carbonate Apatite (Dahlite) (677) Quartz and unidentified substances (357)	❖ and ❖❖ Figs 7, 8 and 58; Textfig. 2a G 1966 a	Grey-brown crystalline Slate-coloured and powdery, golden-brown, iridescent with violet metallic hues, Outer surface yellow.
Orthocone (adapical portion) (420)	*Lower Pennsylvanian* Kendrick Shale Cow Creek, Kentucky U.S.A.	Aragonite & Calcite (weak ray 3.03 Å) Aragonite	❖ np sph	Pink-grey Chalky-powdery
Euloxoceras angustius (1214) (Saunders, 1973)	*Carboniferous* Mississippian Imo Shale, Saunders Locality, Arkansas, U.S.A.	Calcite Aragonite (1 ray)	cyl varic angul	Black Crystalline

The conchiolin matrices in nacreous layers

Reticycloceras peytonense (1213)	*Carboniferous* Mississippian Imo Shale, Saunders Locality, Arkansas, U.S.A.	Calcite	angul	Black-brown Crystalline
Isorthoceras sociale Flower (453)	*Upper Ordovician* Maquoketa Formation Graf, Dubuque County, Iowa, U.S.A.	Calcite	❖ G 1966a	Iridescent, golden-blue Grey powdery
Reticycloceras croneisi (1206)	*Carboniferous* Mississippian Imo Shale, Saunders Locality, Arkansas, U.S.A.	Calcite	angul sph ves	Black-brown Crystalline
Unidentified Orthocerids (693, 694, 708, 709)	*Upper Middle Devonian* Gerolstein and Büdesheim, Eifel, Germany	Iron Sulfide	❖	Dark-mahogany-brown (pyrite)
n. gen., n. sp. (1130) (seems to be closely related to the group of Upper Paleozoic orthoceracones generally referred to *Bactrites*: James L. Murphy, 1970)	*Pennsylvanian* Putnam Hill Limestone Member, Crory Road near Canfield Township, Mahoning County, Ohio, U.S.A.		sph	Pink Chalky

O. ONCOCERIDA.

Rizoceras sp. (1051)	*Upper Devonian* Virgin Hills Formation Bugle Gap, Western Australia	Calcite	❖	White Crystalline
Augustoceras shideleri Flower (1032)	*Upper Ordovician* Cincinnatian Leipers Formation Rowena, Kentucky, U.S.A.	Calcite	ves	Dark-grey Crystalline
Cyrthoceratites sp. cf. *C. depressus* (Bronn) (1071, 1128)	*Upper Middle Devonian* (Givetian) Sötenicher Mulde, Eifel, Germany	Calcite	ves	Dark-brown Crystalline
Archiacoceras, n. sp. (682, 1070, 1082, 1084)	*Upper Middle Devonian* (Givetian) Sötenicher Mulde, Eifel, Germany	Calcite	❖ cyl varic mb an- gul sl sph ves	Dark-brown and black Crystalline
Unidentified Oncocerids (678, 679, 680, 684, 689, 691, 692)	*Upper Middle Devonian* (Givetian) near Gerolstein, Eifel, Germany	Calcite	❖ cyl varic (679) (cf Miocene *Aturia australis*) sph (689) angul	Grey-brown and dark-brown Crystalline

O. DISCOSORIDA

Gomphoceras sp. (711)	*Middle Devonian* near Gerolstein, Eifel, Germany	Calcite	❖❖	Brown Crystalline
Faberoceras sp. (Flower & Teichert, 1957) (697, 699)	*Upper Ordovician* Leipers Beds Cumberland River Valley near Rowena, Kentucky, U.S.A.	Calcite	❖❖ cyl varic angul mb G 1966a, 1967	Dark-brown and grey-brown Crystalline

O. TARPHYCERIDA

Estonioceras imperfectum qu. sp. (791)	*Lower Silurian (Ordovician)* Vaginatenkalk B3 E. K. Kunda, Estland.	Calcite & unid. rays	❖❖ cyl fl broa varic sl	Grey Crystalline
Tragoceras (Planctoceras) falcatum (Schlotheim) (491)	*Silurian (Ordovician)* Sarbja 6, Bromberg, (Bydgosz), Poland	Aragonite Calcite & Aragonite (one ray)	❖	Aragonite: pink chalky material on the inner part of the shell wall. Calcite and aragonite: golden-iridescent powdery substance overlying the internal mould

O. NAUTILIDA

Germanonautilus sp. (1027)	*Middle Triassic* Oberer Muschelkalk Ohrdruff, Thüringen, Germany	Calcite	❖	Black-brown Crystalline
Metacoceras macchesneyi n. sp. Topotype (1138) (James L. Murphy, 1970)	*Carboniferous Pennsylvanian* Brush Creek Limestone Member, Madison Township, Columbiana County, Ohio, U.S.A.	Calcite	sph	Black, crystalline and very hard pink-chalky
Tylonautilus n. sp. (1219)	*Carboniferous Mississippian*, Pitkin age Lick Mt, Newton County, Arkansas, U.S.A.	Calcite		Grey Crystalline
Rutoceratidae, sp. (899)	*Lower Middle Devonian* Gerolstein, Eifel, Germany	Calcite and Aragonite (one ray)	❖	Brown Crystalline
Rutoceratidae sp. (784)	*Lower Middle Devonian* Geeser Schichten, Gees, Eifel, Germany	Calcite	❖ fl sl	Dark-brown Crystalline

The conchiolin matrices in nacreous layers

Specimen	Age / Locality	Mineral	Figure / References	Description
Rutoceratidae sp. (685)	*Upper Middle Devonian* Givetian, Sötenicher Mulde, Eifel, Germany	Calcite	✻ Fig. 19 angul G 1966a	Dark-brown Crystalline
Rutoceratidae sp. (*Cyrtoceras nodosum*) Goldfuss (681)	*Upper Middle Devonian* Givetian, Gees, Gerolstein, Eifel, Germany	Calcite	✻ fl broa G 1972c	Grey-brown and black-brown Crystalline
Gyroconic specimen (*Cyrtoceras?*) (674)	*Lower Middle Devonian* Couvinian Quarry Haine, Couvin, Belgium	Calcite	✻	Dark-grey Crystalline
Large Gyroconic specimen (*Cyrtoceras?*) (713) Service Géologique de Belgique R 144	*Lower Middle Devonian* Couvinian Dinant, Belgium	Calcite Quartz	✻ Fig. 16 Cyl varic fl mb sph ves G 1966a; GVF, 1970; VFG, 1971, 1972.	Hard Coal-black and dark-brown Crystalline
Endolobus clorensis Collinson (969) United States Geological Survey 15075-PC (Gordon, 1964)	*Carboniferous* Mississippian Arkansas U.S.A.	Navajoite	✻ Figs 20, 21, 52 cyl varic fl ves G 1972 c	Grey and black-brown Crystalline

Temnocheilus bellicosus (Morton) (1132) (James L. Murphy, 1970)	*Carboniferous* Pennsylvanian Columbiana Limestone Green Township, Mahoning County, Ohio, U.S.A.	Aragonite	np sph	Grey Crystalline
Stroboceras sp. (1220)	*Carboniferous* Mississippian Pitkin Age, Lick Mnt Newton County Arkansas U.S.A.	Calcite	sph	Grey-brown Crystalline
Domatoceras shepherdi (Sturgeon) (1135) (James L. Murphy, 1970)	*Pennsylvanian* Columbiana Limestone Member, Allegheny Group, near Green Township, Mahoning county, Ohio, U.S.A.	Aragonite	mb sph	Grey Chalky
Domatoceras sp. (*Stearoceras*) (422, 894)	*Permian* San Andres Limestone Rio Penasco River, N.E. of Alamogordo, New Mexico, U.S.A.	Calcite	❋ cyl varic sph ves G1966a, 1968, 1972b, VFG 1971	Grey-brown and brown. Crystalline and golden-iridescent powdery.

The conchiolin matrices in nacreous layers

Solenochilus sp. (1131) (James L. Murphy, 1970)	*Pennsylvanian* Brush Creek Limestone Member, Madison Strip Mine Township Columbiana County, Ohio, U.S.A.	Aragonite	Fig. 45 sph ves	Pink chalky and brown crystalline
Liroceras liratum (Girty) (1137) (James L. Murphy, 1970)	*Carboniferous Pennsylvanian* Kendrick Shale, Cow Creek, Floyd County, Kentucky, U.S.A.	Aragonite	Fig. 46 sph	Pink chalky
Liroceras bicostatum (1208)	*Carboniferous* Mississippian Imo Formation, Arkansas, U.S.A.	Calcite		Black Crystalline
Unidentified Nautilida (749, 751, 752)	*Lower Middle Pennsylvanian* Buckhorn Asphalt Sulphur, Oklahoma, U.S.A.	Aragonite	✽ np cyl varic fl mb G 1966a	Nacreous Violet slabs with bright metallic hues.
Spicatoceras wadei Collins (1030)	*Upper Devonian* Virgin Hills Formation North part of Bugle Gap area. Kimberley, Western Australia	Calcite	✽	White crystalline with red coating

Unidentified *Nautilaceae* (924, 952, 957)	*Lower Jurassic* Lias Lyme Regis & Charmouth, Dorset, England	Calcite	❖	Light-and honey-brown Crystalline
Unidentified Nautilida (963)	*Cretaceous* Westfalen, Germany	Calcite	❖ mb (similar to *Michelinoceras* (1069) and *Aturia australis* (1077)	Very hard, rust-coloured Crystalline
Unidentified Nautilida (990)	*Lower Jurassic* Lias, Whitby, Yorkshire, England	Aragonite	❖❖ fl angul nod VFG 1971; G 1972 b	Hard, chalky Strongly iridescent with metallic hues.
Unidentified Nautilida (323, 335)	*Upper Eocene* Bartonian Wemmelsands, Belgium	Aragonite	Textfig. 6 a np cyl varic sph G 1959 a	Nacreous and chalky
Unidentified Nautilida (1119, 1122, 1123 (30 cm in diameter), 1124, 1125, 1126) (Olsson, 1928)	*Eocene* Lobitos, Piura, Peru.	Calcite	❖ Figs 3, 22, 23 Cyl varic angul mb sph	Brown crystalline (1119, 1122), Honey-brown crystalline (1123, 1124) Yellow-brown crystalline (1125)
Nautilus imperialis J. Sowerby (536) Sedgwick Museum, (Cambridge, England: C. 21. 571)	*Eocene* London Clay Portsmouth, Hantshire, England	Aragonite Unidentified ray (4.33 Å)	❖ sph GL 1972	Brown iridescent Crystalline

Species	Age / Locality	Mineralogy	Symbols	Description
Nautilus sp. (555)	*Upper Eocene* Plainberg im Becken von Reichenhall-Salzburg, Austria	Aragonite	❖ cyl varic angul mb sph G 1968	Light brown Nacreous
Cenoceras sp. (835)	*Middle Jurassic* Lower Oolite, England	Calcite (1ray) Calcite & Iron sulfide	❖❖	Blue-grey crystalline Rust-coloured areas Inner whorls: dark-brown crystalline.
Cenoceras sp. (948; 949; 950) 3 unidentified sp.	*Lower Jurassic* Lower Lias, region of Lyme Regis Charmouth, Dorset, England	Calcite (948, 950) Aragonite & Calcite (949)	❖ to ❖❖ Fig. 47 np angul sph GFV 1970; VFG 1971	948: Honey-brown crystalline. 949: Chalky powdery, iridescent layers. Hard nacreous iridescent layers. Very hard, honey-brown crystalline 950: Hard, honey-brown.
Eutrephoceras dekayi Morton (359)	*Upper Cretaceous* Senonian, Pierre Shale Formation of the Montana Group, Black Hills, South Dakota, U.S.A.	Aragonite 1 ray of Calcite (3.03 Å)		Hard Grey-brown Nacreous
Eutrephoceras sp. (366) (T. Birkelund, 1965: 8811)	*Upper Cretaceous* Senonian (Maestrichtian?) The Agatdalen Núgssuaq New Oyster Ammonite Locality, West Greenland	Aragonite & Calcite (very weak ray 3.03 Å)	❖ np cyl varic	Brittle, nacreous, strongly iridescent with metallic hues.

Eutrephoceras sp. (398, 999) United States Geological Survey 27065	*Cretaceous* Eutaw Formation, Menabites, Alabama, U.S.A.	Aragonite	❖ Fig. 24 VFG 1971; G 1972 c	Chalky iridescent with metallic hues. 999: hard.
Eutrephoceras victorianum (Teichert, 1943) (1081)	*Middle Paleocene* Pebble Point Formation N. W. Side Point Pember S. E. Princetown, Victoria, Australia	Aragonite	❖❖ fl broa sl	Nacreous, violet flakes with strong metallic hues.
Eutrephoceras balcombensis (Chapman, 1915) (1076)	*Middle Miocene* Fyansford Formation Balcombian Heatherton, Victoria, Australia	Aragonite	❖ mb GVF 1970; VFG 1971	Hard White chalky Inner part of nacreous layer iridescent with metallic hues.
Cymatoceras sp. (478)	*Cretaceous* Gault, Albian, Vöhrum, Hannover, Germany	Aragonite	❖ fl sph VFG 1971	Nacreous iridescent Inner layer pink chalky powdery
Sp. near *Cymatoceras* (972)	*Lower Jurassic* Lias Area of Lyme Regis Charmouth, Dorset, England	Calcite	❖	Dark-brown crystalline
Aturoidea distans Teichert 1943 (1080)	*Middle Paleocene* Pebble Point Formation Buckleys Point, S. E. of Princetown, Victoria, Australia	Aragonite	❖ Fig. 26 np(swollen)	Yellow-orange-coloured Crystalline

The conchiolin matrices in nacreous layers

Aturia sp. (926)	*Paleocene-Miocene* probably Eocene (England) origin unknown	Aragonite	❖ Fig. 27	Nacreous
Aturia zic-zac SOW. (407)	*Eocene* near Torino, Italy	Aragonite & Calcite (ray 3.03 Å)	❖ fl sl nod	Nacreous, iridescent
Aturia sp. (477)	*Lower Eocene* Island of Sheppey, England	Aragonite & Calcite (Ray 3.03 Å)	❖	Pink-silvery Nacreous
Aturia clarkei Teichert 1944 (1078)	*Upper Eocene* Aldingan Browns Creek Clay near mouth of the Johanna River, Johanna, S. W. Victoria, Australia	Aragonite	❖❖ rib	White yellow Nacreous and chalky
Aturia sp. ? (1022, 1048)	*Eocene* Wemmelian, Baeleghem, Flanders, Belgium	Aragonite	❖ Fig. 25 np fl cyl varic mb ves VFG 1971, 1972	1022: very hard, snow-white Chalky, faintly iridescent. 1048: very brittle, snow-white powdery, faintly iridescent.
Aturia sp. (333, 344, 353, 355)	*Middle Oligocene* Upper Rupelian Boom (clay-pits), Belgium	Aragonite	np cyl varic fl sph G 1959a	Iridescent Nacreous

Aturia australis McCoy 1876 (1079)	*Upper Oligocene* Jan Juc Formation Janjukian Cliff at Fischerman's steps, S. W. of Torquay, Victoria, Australia	Aragonite	❖ sph	Nacreous
Aturia australis McCoy 1876 (535) Sedgwick Museum, Cambridge, England, C 5341	*Oligocene-Miocene* prob. Balcombian Muddy Creek, Victoria, Australia	Aragonite	❖ np	Light brown
Aturia australis McCoy 1876 (1077)	*Middle Miocene* Muddy Creek Formation Balcombian Clifton Bank, Muddy Creek, Hamilton, Western Victoria, Australia	Aragonite	❖❖ swollen np, fl broa cyl varic mb sph	Nacreous
Aturia luculuensis (608) U.S.N.M.	*Miocene* Quipai, Sea Cliffs, Massandombe, Angola	Aragonite	❖ nod mb	Ivory-coloured Crystalline
Aturia aturi (Coll. Mayer-Eymar-Gubler) (1252)	*Miocene* Aquitaine Locality unknown, France		cyl varic sph sl scarce, poorly preserved conchiolin	

Aturia aturi Basterot (Coll. Mayer-Eymar) (1251)	*Miocene* Laughian Saucats near Bordeaux, France	mb sph scarce, poorly preserved conchiolin	
Unidentified nautiloid (septum) USGS Locality 25885 (1254)	*Upper Oligocene* Nye Formation Yaquima Quadrangle in float, north of Mouth of Grant Creek, Lincoln County, Oregon, U.S.A.	sph scarce, poorly preserved conchiolin	Nacreous Brown

Table 2: PALEOZOIC AMMONOIDS (1)

SUBORDER ANARCESTINA

Agoniatites vanuxemi Hall (704)	*Middle Devonian* Cherry Valley Limestone, Stockbridge Falls, New York, U.S.A.	Calcite	❖ np(?) nod ves sph G 1968	Dark-brown & Coal-black Crystalline
Manticoceras sp. (1031)	*Upper Devonian* Virgin Hills Formation, Bugle Gap Western Australia	Calcite	❖❖ mb	White (red coating) Crystalline

SUBORDER CLYMENIINA

Clymenia s. str., (cf. *striata* Münst.) (790)	*Upper Devonian* Burg bei Rösenbeck, Germany		❖ mb sl	Dark-brown
Prob. Clymeniaceae (695)	*Upper Devonian* Büdesheim, Eifel, Germany	Iron sulfide	❖ Textfig. 2b fl Sph mb sl	Bright dark-brown Pyritized

SUBORDER GONIATITINA

Unidentified Goniatites (513, 515)	*Upper Devonian* Büdesheim, Eifel, Germany	Iron sulfide	❖ varic G 1972b	Mahogany-brown Crystalline

The conchiolin matrices in nacreous layers

Species	Age/Locality	Mineral	Notes	Appearance
Tornoceras ausavense (Steininger) (1072)	*Upper Devonian* Upper Frasnian, Mariembourg, Belgium	Calcite	bro sl	Black Crystalline
Tornoceras sp. (1073, 1074)	*Upper Devonian* Mariembourg & Couvin, Belgium	Calcite	❖❖ (1073) Sl ❖ (1074) bro fl sl	Black Crystalline
Tumulites varians (1207)	*Carboniferous* Upper Mississippian (Chesterian), Fayetteville Shale, Fayetteville, Arkansas, USA	Calcite	Figs 29 & 30 angul	Black Crystalline
Cravenoceras friscoense (1204)	*Upper Carboniferous* Mississippian, Imo Shale Saunders Locality, Arkansas, USA	α-Quartz	mb sl	Black-Brown Crystalline
Cravenoceras fayettevillea (1211)	*Carboniferous* Mississippian, Fayetteville Shale, Fayetteville, Arkansas, USA	Calcite	Fig. 31 angul	Black-brown Crystalline
Cravenoceras hesperium Gordon (1209)	*Carboniferous* Mississippian, Pitkin Limestone Lick Mt, Newton County, Arkansas, USA			Grey-brown Crystalline

Cravenoceras n. sp. (1221)	*Carboniferous* Mississippian, Pitkin Limestone, Lick Mt, Newton county Arkansas, USA	Calcite	Sph	Grey-brown Crystalline
Homoceras beyrichianum (De Koninck) (519, 668, 673)	*Upper Carboniferous* Etage de Chokier-Flémalle, Belgium	Calcite	❖ ves G 1966a	Dark-brown Crystalline
Homoceras undulatum (663) (7 sms)	*Upper Carboniferous* Coolagh River, Lisdoonvarna, County of Clare, Ireland	Calcite	❖❖	Black crystalline
Homoceras sp. (*beyrichianum*?) (672–673)	*Carboniferous* Roadford, Toomullin, County of Clare, Ireland	Calcite	varic angul G 1966a	Black crystalline
Anthracoceras discus (1205)	*Carboniferous* Mississippian, Imo Shale, Saunders locality, Arkansas, U.S.A.	Quartz	mb sl	Black-brown Crystalline
Anthracoceras wanlessi (Plummer & Scott, 1937) (855)	*Middle Pennsylvanian* Carbondale Group, near Lewiston, Illinois, USA	Calcite	❖❖ cyl varic angul sph G 1966a	Brown Crystalline

The conchiolin matrices in nacreous layers

Species	Formation/Locality	Mineralogy	Notes	Appearance
Beyrichoceratoides sp. (662) H. M. Geological Survey of Scotland, Edinburgh M 2135 L	*Lower Carboniferous of Scotland* Upper Oil Shale Group, Cot Castle Shell Bed, Bore Baads Mine, Scotland,	Aragonite (cf. Hallam & O'Hara, 1962)	❋ cyl varic Sph mb G 1966a	Pink-creamy & brown Nacreous
Rhadinites miseri (1212)	*Carboniferous* Upper Mississippian, (Chesterian) Imo Formation, Arkansas, USA	α-Quartz		Black-brown Crystalline
Hudsonoceras proteum (675)	*Carboniferous* Rough Lee, Blackburn, Lancashire, England	Calcite	❋❋ Textfig. 4	Black-brown Crystalline (675)
Hudsonoceras proteum (664, 665) ⟨34 spms⟩	*Upper Carboniferous* Coolagh River Lisdoonvarna, County of Clare, Ireland (664)	Calcite	664: cyl varic angulves mb.	Black and black-brown Crystalline (664–665)
	Exdale Shales, Manitor Derbyshire, England (665)		G 1966a	
Bisatoceras secundum (1248)	*Carboniferous* Pennsylvanian Brentwood Limestone, Bloyd Formation, Gaither Mountain Mc Caleb's Locality N° 1, Arkansas, USA	Calcite	❋ mb nod sl	Brown Crystalline

Species	Formation/Location	Mineral	Symbols	Description
Eoasianites sp. (358)	*Lower Middle Pennsylvanian* Buckhorn Asphalt, Sulphur, Oklahoma, USA	Aragonite & Calcite (Ray 3.03 Å)	❋❋ ves	Bright, smoke-brown Crystalline
Eoasianites hyattianus (Girty) (676) (768) S. U. Iowa, Fig. spm 8.8888	*Lower Middle Pennsylvanian* Buckhorn Asphalt, Sulphur, Oklahoma, USA	Aragonite & Calcite (ray 3.03 Å)	❋❋ np cyl varic angul mb sl ves sph GT 1965; G 1966[a]	Bright, smoke-brown & Dark-brown Crystalline
Gastrioceras circumnodosum group (Roofballs) (666)	*Upper Carboniferous* Lower Westphalian, Coal Mine Soumagne-Micheroux, Belgium		❋❋ Textfig. 2c varic ves nod	Black crystalline
Gastrioceras circumnodosum (listeri? Calver?) (667)	*Upper Carboniferous* Lower Westphalian, Lancashire and N. W. Yorkshire, England		❋ Textfig. 2c angul ves sph	Black crystalline
Arkanites relictum (1249)	*Lower Pennsylvanian* Morrowan, Bradshaw Mountain, Arkansas, USA	Calcite	❋ sl mb	Grey-brown Crystalline
Goniatites sp. (2 spms) (669)	*Carboniferous of Colomb* Bechar, Algeria	Quartz	❋	Red-brown

30 Paleozoic ammonites, mostly goniatites (385, 390, 421), including probably	*Lower Middle Pennsylvanian* Buckhorn Asphalt, Sulphur, Oklahoma, USA	Aragonite Aragonite & Calcite (ray 3.03 Å)	Fig. 14 am p (?) varic fl mb nod ves sph	Bright-Smoke-brown iridescent
Eupleuroceras bellulum Miller & Cline 1934 and several species of *Gastrioceras* (720)			n.p. GGL 1964; GT 1965; G 1966ab, 1972; GL 1972	Dark-brown (720)
Wellerites mohri Plummer & Scott (769) S.U. Iowa 8891 (Miller & Furnish, 1958)	*Lower Middle Pennsylvanian* Buckhorn Asphalt, Sulphur, Oklahoma, USA	Aragonite & Calcite	Fig.1 G 1966a, 1972b	Dark-mahogany-brown & strongly iridescent Nacreous
Pygmaeoceras morrowense Gordon 1960, (*Gaitherites solidum*) (1210)	*Lower Pennsylvanian* Brentwood Limestone, Bloyd Formation Gaither Mountain, McCaleb's Locality n° 11, Arkansas, USA	Iron sulfide Calcite	sl	Light-grey-brown crystalline
Texoxeras texanum (Girty 1908) (858)	late *Lower Permian* Bone Spring Limestone, Culberson County, Texas, USA	Calcite	❖❖ Fig. 48 ves sph	Dark-brown crystalline
Metadimorphoceras sp. (1218)	*Lower Pennsylvanian* Bloyd Formation, Lick Mountain, Newton County Arkansas, USA	Calcite	angul	Grey-brown crystalline

Eothalassoceras n. sp. (1134) (J. L. Murphy, 1970)	*Pennsylvanian* Putnam Hill Limestone, Crory Road. Canfield Township, Mahoning County, Ohio, USA	Aragonite	angul sph	Pink Chalky

SUBORDER PROLECANITINA

Pronorites sp. (754, 750, 1158, 1187) Geol. Survey Univ. Kansas 3771	*Pennsylvanian* Buckhorn Asphalt, Sulphur, Oklahoma, USA	Aragonite Aragonite & Calcite	❖ Fig. 33 am p (?) septal np (1158) cyl varic angul mb sph G 1968, 1966a	Brown (asphalt)
Unidentified spm. (septal material: saddles, lobes) (755) Geol. Survey Univ. Kansas 3768	*Pennsylvanian* Buckhorn Asphalt, Sulphur, Oklahoma, USA	Aragonite & Calcite (ray 3.03 Å)	❖❖ septal np mb G 1966a	Brown (asphalt)
5 *Unidentified goniatites* (septal Material) (420)	*Lower Pennsylvanian* Kendrick Shale Cow Creek, Kentucky, USA	Aragonite & Calcite (ray 3.03 Å)		Pink-grey powdery

Table 3: MESOZOIC AMMONOIDS (1)

SUPERFAMILY 1: OTOCERATACEAE

Species	Age/Locality	Mineralogy	Figures/Notes	Color/Structure
Paraceltites elegans Girty (856)	*late Lower Permian* Bone Spring Limestone, Culberson County, Texas USA	Calcite	❖❖ Figs 32, 49 ves sph	Dark-brown Crystalline
Ophiceras commune Spath (902)	*Lower Triassic* Lower Scythian, Cape Stosch, East Greenland (74° N. lat.)	Calcite Calcite & Aragonite (one ray)	❖❖ nod sph G. 1972c	Brown Crystalline

SUPERFAMILY 2: NORITACEAE

Species	Age/Locality	Mineralogy	Figures/Notes	Color/Structure
Flemingites glaber Waagen (789)	*Triassic*	Calcite	❖ nod sl	Light-brown Crystalline
Meekoceras meetlingizone (788)	*Triassic* Schalshai Cliff, Himalaya	Calcite	❖ cyl varic ves	Black-brown Crystalline

SUPERFAMILY 3: CERATITACEAE

Species	Age/Locality	Mineralogy	Figures/Notes	Color/Structure
Carnites floridus (Wulfen) (965)	*Upper Triassic* Lumachelle, Ceratitenschichten, Kraith bei Bleiberg, Carinthia, Austria	Aragonite	❖❖ Fig. 35 sl sph	Loosely curved slabs, Brown iridescent with strong metallic hues (flame, green)

Ceratitaceae sp. (1026)	*Jurassic* Toarcian, Italy	Calcite	✲ Fig. 34 sl sph	Dark-grey Crystalline

SUPERFAMILY 7: ARCESTACEAE

Joannites klipsteini (E. von Mojsisovics) (1143, 1157)	*Triassic* Carnian, Greece	Calcite	angul	Grey crystalline White crystalline Brown coating

SUPERFAMILY 10: PHYLLOCERATACEAE

Rhacophyllites neojurensis Quenstedt (794)	*Triassic*	Calcite	✲✲ mb nod sl sph	Brown Crystalline
Phylloceratidae sp. (825)	*Lower Jurassic* Upper Lias, Sinemurian, Whitby, Yorkshire, England	Calcite	✲ fl	Black-brown Crystalline
Phylloceratidae sp. (1156) (Buccelle & Garavello, 1967 ab)	*Cretaceous* Aptian/Albian, La Stua, Cortina d'Ampezzo, Italy	Calcite	✲	Grey crystalline
Tragophylloceras loscombi (J. Sowerby) (839)	*Lower Jurassic* Lower Lias, Dorset, England	Calcite	✲ angul	Brown crystalline

Phylloceras sp. (1193, 1199) (14 specimens)	*Jurassic* Oxfordian Col de Perruergues, Drôme, France	Calcite	angul	1193: Blue-brown 1199: Dark-brown Crystalline

SUPERFAMILY 11: LYTOCERATACEAE

Lytoceras sp. (489)	*Lower Middle Jurassic* Boll, Württemberg, Germany	Aragonite		Pink Chalky
Lytoceras jurense (septum) Quenstedt (964)	*Lower Jurassic* Middle Lias (Pliensbachian) Reutlingen, Württemberg, Germany	Calcite	❖❖ Sph	Dark-brown Crystalline
Lytoceras sp. (1025)	*Jurassic* Domerian–Toarcian, Provincia di Bergamo, Italy	Calcite	❖❖ angul	Red-brown crystalline
Ammonites lineatus penicillatus (785)	*Jurassic* Brauner Jura alpha, Gammelshausen bei Boll, Württemberg, Germany	Aragonite & Calcite	❖ Figs 4, 5, 6 mb GL 1972; G 1972bc. VFG 1972	Hard Grey- and dark-brown iridescent Crystalline
Saghalinites sp. (1159)	*Cretaceous* Maestrichtian, The Agatdalen, Núgssuaq, West Greenland	Aragonite	fl sph	Strongly iridescent with metallic hues Nacreous

SUPERFAMILY 14: TURRILITACEAE

Cirroceras bornbyense (Whiteeaves) (551)	*Uppermost Cretaceous* Maestrichtian-Nigerian		❖	Chalky
Hamites sp. (?) (518)	*Cretaceous* Greensand, Cambridge, England	Aragonite	sph mb	Iridescent Chalky
Baculites ovatus Say (356)	*Upper Cretaceous* Senonian, Pierre Shale of Montana Group, Colorado Springs, Colorado, USA	Aragonite	❖ sph G 1972b	Strongly iridescent with metallic hues Nacreous
Baculites sp. U.S. Geological Survey, 18.756 (387)	*Cretaceous* Claggett Shale, West flank of Grouse-Alder Dome Montana, USA	Aragonite and ray 3.03 Å of Calcite	❖❖ pb mb	Yellow-brown with violet hues Nacreous Iridescent
Baculites sp. U.S. Geological Survey, 10.959 (392)	*Cretaceous* Bear Paw Shale, Fort Belknap Reservation, Montana, USA	Aragonite	sph	Pink-Silvery Iridescent Nacreous

Baculites claviformis Stephenson U.S. Geological Survey, 25 406 (393, 553, 998, 395)	*Upper Cretaceous* Senonian, Ripley Formation Coon Creek Member, Coon Creek, near Enville-Adamsville, McNairy County, Tennessee, USA	Aragonite	❖ fl cyl varic ves sph mb (393–998) mb (553), sph mb (395) G. 1966a; 1972b; GVF, 1970; VFG 1971; GL 1972	393, 395 strongly iridescent with metallic hues 553 nacreous 998: white-creamy Brittle Chalky
Baculites obtusus Meek U.S. Geological Survey Mesozoic Locality D 2624 (1002)	*Upper Cretaceous* Lower part of Claggett Shale Lower part of Upper Campanian Rosebudy County, Montana, USA	Aragonite Aragonite & Calcite (one ray)	❖	Brown-glassy & Pink-white Nacreous
Baculites scotti Cobban U.S. Geological Survey, Mesozoic Locality D 903 (1003)	*Upper Cretaceous* Lower part of Pierre Shale, Middle part of Upper Campanian, Fall River County, South Dakota, USA	Aragonite	❖ ves sph	Strongly iridescent Nacreous
Baculites compressus Say (480, 486, 502, 503, 504, 506b, 508, 509)	*Cretaceous* Senonian, Bad Lands of Dakota, USA	Aragonite (480, 504) Calcite (ray 3.03Å) and Iron Sulfide (502) Aragonite & Calcite (ray 3.03 Å) (509)	❖	Strongly iridescent Nacreous (480, 486, 509) Pink, Chalky (502) Brown, Chalky & Nacreous (503, 504)

SUPERFAMILY 15: SCAPHITACEAE

Species	Age/Locality	Mineralogy	Figs/Notes	Appearance
Scaphites rosenkrantzi n. sp. (T. Birkelund, 1965, Pl. 22) (364–365)	*Upper Cretaceous* Senonian/Campanian, Scaphites naesen, The Agatdalen Núgssuaq, West Greenland	Gypsum Aragonite & Calcite	❖ Figs 36, 37; Textfig. 5c am p(?) np(?) nod GL 1972	Strongly iridescent with metallic hues (blue, green, violet) Nacreous
Scaphites sp. Kansas Geological Survey 11775 (434)	*Cretaceous* Gulfian, Blue Hills Shale Member, Carlile Shale, Kansas, USA	Aragonite & Calcite (one ray)		Iridescent Pink-white Nacreous
Scaphites nodosus Say U.S. Geological Survey 19003 (394, 505)	*Cretaceous* Black Hills, South Dakota, USA (394)	Aragonite & Calcite (ray 3.03 Å)	Sph (394)	Iridescent Creamy (505) Pink-creamy (485)
Acanthoscaphites sp. (*nodo sus* Say?) (485)	*Upper Cretaceous* Senonian, Bad Lands of Dakota USA (505, 485)			Nacreous
Discoscaphites n. sp. (7 spms) (T. Birkelund, 1959) (363)	*Upper Cretaceous* Maestrichtian, "Oyster Ammonite Locality" The Agatdalen Núgssuaq, West Greenland	Aragonite		Nacreous Strongly iridescent with metallic hues

The conchiolin matrices in nacreous layers 113

			Fig. 38 n p nod	Iridescent
Discoscaphites sp. U.S. Geological Survey 21 740 (388)	*Cretaceous* Fox Hills S S, Little Eagle, Carson County, South Dakota, USA	Aragonite		Nacreous & Chalky
Discoscaphites sp. U.S. Geological Survey 9 (1882) (397)	*Cretaceous* Right border of Yellowstone Valley, near Glendive, Montana, USA	Aragonite & Calcite (ray 3.03 Å)	Abundant	Strongly iridescent with intense metallic hues Nacreous
Discoscaphites sp. U.S. Geological Survey 5879 (400)	*Cretaceous* Fox Hills S.S. Cheyenne Indian Reservation. Bear Creek, South Dakota, USA	Aragonite	Abundant conchiolin; sph mb	Strongly iridescent (metallic hues) Nacreous
Discoscaphites sp. U.S. Geological Survey 10771 (401)	*Cretaceous* Bear Paw Shale, Concretions in Bed of Prairie Elk Creek, Montana, USA	Calcite & Aragonite (one ray)		Strongly iridescent (pink-copper hues) Nacreous
Hoploscaphites sp. U.S. Geological Survey Mesozoic Locality 23625 (1004)	*Upper Cretaceous* Upper part of Bear Paw Shale, Lower Maestrichtian, Richland County, Montana USA	Aragonite	❖	Strongly iridescent with metallic hues Powdery

Hoploscaphites nicolleti (Morton) U.S. Geological Survey Mesozoic Locality 21740 (1005) (1201)	*Upper Cretaceous* Middle part of Maestrichtian, Carson County, South Dakota, USA (1005) Fox Hills Formation (1201)	Aragonite Calcite (1201: innermost shell wall)	❋ fl broad angul mb sph GL 1972	Strongly iridescent Nacreous

SUPERFAMILY 16: PSILOCERATACEAE

Psiloceras planorbis (J. de C. Sowerby) (567, 988)	*Lower Jurassic* Lower Lias, Hettangian, Planorbis Zone Range, Watchet, Somerset coast, England	Aragonite (988) Aragonite & Calcite (ray 3.03 Å) (567)	❋❋ Fig. 2 mb sph VFG 1971; G 1972b	Considerably flattened, brittle specimens Strongly iridescent with brown-violet hues Nacreous, glassy, dark-brown flakes,
Arietites brooki (J. Sowerby) (539) Sedgwick Museum, J. 42809	*Lower Jurassic* Lower Lias, Lyme Regis, Dorset, England	Calcite	❋❋ sph	Light-honey-brown, lustreless, crystalline.
Arietites obtusus (J. Sowerby) (540) Sedgwick Museum J 42812	*Lower Jurassic* Upper Lias, Whitby, Yorkshire, England	Calcite	❋❋ sph	Light-honey-brown Crystalline
Oxynoticeras sp. (942) (*Fastigiceras?*)	*Lower Jurassic* Lower Lias, Lyme Regis, Dorset, England	Calcite	❋❋ VFG 1971	Hard Honey-brown and white Crystalline

The conchiolin matrices in nacreous layers 115

Asteroceras obtusum (Sowerby) (916)	*Lower Jurassic* Lower Lias, Charmouth, Dorset, England	Calcite	✱✱ fl	Honey-brown crystalline
Asteroceras sp. 12 large specimens (diameter 40 cm: 941) (945, 921, 967, 930, 941, 946, 973, 1160, 1161, 1174)	*Lower Jurassic* Lower Lias, Lyme Regis and Charmouth, Dorset, England	Calcite	✱✱ Cyl varic (1161) mb ves sph 1161: Curved segments of trabeculae GVF 1970; VFG (941) 1971	Hard Honey-brown crystalline Thin iridescent, chalky inner (973) and outer (921) layers. 941: brittle iridescent, pink-creamy and hard honey-brown layers. 1160: strongly iridescent, nacreous and honey-brown crystalline.
Asteroceras stellare (J. Sowerby) (929)	*Lower Jurassic* Lower Lias, Lyme Regis and Charmouth, Dorset, England	Calcite	Figs 39 and 40 Cyl	Dark brown Crystalline
Echioceras sp. (from boulders) (928, 1179)	*Lower Jurassic* Lower Lias, Charmouth, Dorset, England	Calcite	am p(?) sph	Brown Crystalline
Asteroceras? Echioceras? (910)	*Lower Jurassic* Lower Lias, Cliffs of Charmouth, Dorset, England	Calcite	✱ fl broa angul mb sph	Honey-brown Crystalline

SUPERFAMILY 17: EODEROCERATACEAE

Microderoceras birchii (1162)	*Lower Jurassic* Lower Lias, Lyme Regis, Dorset, England	Calcite	cyl varic	Grey-brown Crystalline
Microderoceras sp. (843, 845)	*Lower Jurassic* Lower Lias, Lyme Regis, Dorset, England	Calcite	✻ sph	Light-brown Crystalline
Promicroceras marstonense (Spath) (823)	*Lower Jurassic* Lower Lias, Ammonite Marble, Marston Magna, Yeovil, Somerset, England	Calcite	✻ fl G 1968	Dark-brown Crystalline
Eoderoceras armatum (833)	*Lower Jurassic* Lower Lias, Dorset, England	Iron sulfide & Calcite (one ray)	✻ fl cyl mb nod	Black-brown Crystalline Strongly pyritized
Liparoceras sp. (986)	*Jurassic* Lias, Snows Hill, Gloucestershire, England	Calcite	✻ fl angul mb very abundant	Pink Chalky
Androgynoceras sp. (828) (830)	*Lower Jurassic* Lower Lias, Lyme Regis, England	Calcite	✻ fl cyl varic ves (830) G. 1968; GL 1972	Dark-brown Crystalline

The conchiolin matrices in nacreous layers 117

Androgynoceras lataecosta (837)	*Lower Jurassic* Lower Lias, Cotswolds, England	Aragonite & Calcite	❖ fl mb	Pink-white Chalky
Amaltheus spinatus Bruguière *var. spinata Br., stadi. Costatum* Rein. (408)	*Lower Jurassic* Middle Lias, Schloss Banz, Franken, Germany	Aragonite	❖ Fig. 41 n p(?) sph G 1968	Blue iridescent, nacreous, & pink-creamy Chalky
Amaltheus (Pleuroceras) costatus Bruguière (970)	*Lower Jurassic* Middle Lias, Domerian, Quedlinburg, Harz, Germany	Aragonite	❖❖ ves	Pink Chalky
Amaltheus margaritatus De Montfort (795)	*Lower Jurassic* Lias delta, Algernissen, North Germany	Aragonite & Calcite	❖ abundant	Pink-iridescent Chalky
Amaltheus sp. (956, 1163)	*Lower Jurassic* Lyme Regis, Dorset, England	Calcite		Ivory-yellow-coloured Crystalline
Amaltheus sp. (987)	*Jurassic* Oxford Clay, Weymouth, Dorset England	Calcite	❖ mb VFG 1972	Dark-brown Crystalline

Pseudamaltheus engelhardti (d'Orb.) (962)	*Lower Jurassic* Middle Lias, Domerian, Dörnten bei Goslar, Harz, Germany	Aragonite Iron Sulfide	❖❖ fl cyl varic sph mb GVF 1970; VFG 1971	Very brittle Pink chalky Iridescent (metallic hues)
Pleuroceras spinatum (Bruguière) (501)	*Lower Jurassic* Middle Lias, Schloss Banz, Franken, Bayern, Germany	Aragonite	❖ fl broa G 1968; VFG 1971	Brittle Pink creamy & Blue iridescent Chalky
Pleuroceras spinatum (Bruguière) (827, 829, 838, 840)	*Lower Jurassic* Middle Lias, Southern England and Yorkshire (838), England	Calcite Calcite & 2 unid. rays (829)	❖❖ fl broa (827) angul (82-9) very abundant (838)	Dark-brown Crystalline Rust-brown & Grey (840)
Dactylioceras sp. (339-343 and 345)	*Lower Jurassic* Lower Toarcian, Dontrieu, Marne, France	Aragonite	❖ G 1959	Strongly iridescent Nacreous
Dactylioceras commune (J. Sowerby) (522-2)	*Lower Jurassic* Lias, Bifrons Zone, Commune Subzone, Whitby, Yorkshire England	Calcite	❖❖ cyl varic angul	Dark-brown Crystalline
Porpoceras sp. (834)	*Lower Jurassic* Upper Lias, Yorkshire, England	Calcite	❖❖ fl broa very abundant	Brown Crystalline

SUPERFAMILY 18: HILDOCERATACEAE

Harpoceras mulgravium (Simpson, Young & Bird) (523, 524, 452)	*Lower Jurassic* Upper Lias, Falcifarum Subzone, The Scars, Whitby, Yorkshire, England	Calcite Calcite, Iron Sulfide & Quartz (523) Calcite, Iron Sulfide & unid. substances (524) Calcite & Aragonite (ray 3.36Å) (452)	❖❖❖ cyl angul ves G 1966a, 1967, 1968; GL 1972	Dark-honey-brown (calcite) Golden-coloured (Iron Sulfide) Crystalline
Harpoceras exaratum (Young & Bird) (538) Sedgwick Museum J 42575	*Lower Jurassic* Upper Lias, Whitby, Yorkshire, England	Calcite		Black-brown Crystalline
Harpoceras falcifarum (J. Sowerby) (971)	*Lower Jurassic* Upper Lias, Toarcian, Falcifarum Zone, Ilminster, England	Calcite	❖❖	Dark-grey Crystalline
Harpoceras sp. (1197)	*Lower Jurassic* Toarcian, La Verspillière, Isère, France	Calcite		Dark-rusty-brown
Eleganticeras sp. (832)	*Lower Jurassic* Upper Lias, Whitby, England	Calcite	❖❖ angul G. 1968	Dark-brown Crystalline
Ovaticeras sp. (831)	*Lower Jurassic* Upper Lias, England	Calcite Calcite & BaSO$_4$	❖ fl broa	Chocolate-brown Crystalline

Hildoceras sp. (824, 826, 1009)	*Lower Jurassic* Upper Lias, Whitby and other localities Yorkshire, England	Calcite	❊ cyl varic angul mb ❊ ❊	Grey-black. Crystalline (824) Dark-brown. Crystalline
Hildoceras sp. (1008)	*Lower Jurassic* Lower Lias, Lyme Regis, Dorset, England	Calcite	❊ ❊ Figs 42, 51; Textfig. 5b Cyl varic sph ves angul very abundant G 1972b	Dark-brown Crystalline
Staufenia sinon (Bayle) (799) (Orig. Rieber, 1963-CE 1211–115 Tübingen)	*Jurassic* Dogger beta, Scheffen (Wutach), Germany	Calcite	❊ ❊	Brown Crystalline
Leioceras opalinum Reinecke (362, 367)	*Middle Jurassic* Upper Aalenian, Opalinus-Ton, Osterfeld Ziegelei, Tongruben, Goslar, Harz, Germany	Aragonite	❊ am p (?) n p(?) GGL 1964; G 1972b	Hard 362: strongly iridescent with intense metallic hues
Leioceras (*Harpoceras*) *opalinum* (Rein.) (405) 10 spms	*Middle Jurassic* Upper Aalenian, Lower Dogger Opalinus-Ton, Boll, Württemberg, Germany	Aragonite & ray 3.03 Å of Calcite	❊ cyl varic nod sph GGL 1964; G 1968	Brittle Grey-White, Pink-White chalky-powdery

The conchiolin matrices in nacreous layers 121

Species	Age/Location	Mineralogy	Notes	Description
Leioceras opalinum Reinecke (494, 966)	*Middle Jurassic* Lower Aalenian, Harz, Germany	Aragonite	❊❊ G 1972b; GL 1972	Pink-White Chalky, powdery
Ludwigia murchisonae (836)	*Lower Jurassic* Lias, Lower Oolite, Dorset, England	Calcite	❊❊ angul G 1972b	Brown Crystalline
prob. *Graphoceratidae* (*Ludwigia* sp.) (842)	*Jurassic* Lower Oolite, Dorset, England	Calcite	❊❊ angul	Grey-rust-brown Crystalline

SUPERFAMILY 19: HAPLOCERATACEAE

Species	Age/Location	Mineralogy	Notes	Description
Oppelia sp. (841)	*Jurassic* Lower Oolite, England	Calcite	❊ mb	Grey and rust-brown Crystalline
Neumayria catenulata Neum. (*Taramelliceras*) (514)	*Upper Jurassic* Choroschowo, U.S.S.R.	Dahlite Quartz	❊	Strongly iridescent with smoke-brown metallic hues. Chalky

SUPERFAMILY 20: STEPHANOCERATACEAE

Species	Age/Location	Mineralogy	Notes	Description
Stepheoceras sp. (*Stephanoceras*) (989)	*Lower Jurassic* Lias, Whitby, Yorkshire, England	Calcite	❊❊ Fig. 43, Textfig. 5a angul sph abundant G 1972c	Dark-chocolate-brown Crystalline

Kosmoceras Jason Reinecke (471) *Kosmoceras* sp. (Jason n° 2) (Reinecke) (473) *Ammonite* Jason Zone (472)	*Upper Jurassic* Lowest zone of Lower Oxfordian (Jason Zone), Pits of the London Brick Co., Peterborough, England	Aragonite Aragonite Aragonite	❖ (471) ❖❖❖ (472) nod ❖❖ (473)	Strongly iridescent Pink chalky Powdery (472) Nacreous, creamy (473)
Gulielmiceras (Ammonites gulielmi) (J. Sowerby) (475)	*Upper Jurassic* Lower Zones of Lower Oxfordian, Pits of the London Brick Co., Peterborough, England	Aragonite & Calcite	❖ nod	Pink- Chalky
Cadoceras elatmae (796, 797)	*Jurassic* Callovian Elatma?	Aragonite & Calcite	❖❖ sph mb	Strongly iridescent Nacreous & chalky
Quenstedtoceras (Lamberticeras) lamberti (Sowerby) (466)	*Upper Jurassic* Lower Zones of Upper, Oxfordian clay, Pits of the London Brick Co., Peterborough, England	Calcite & Iron sulfide	❖ sl	Pink-chalky
Cardioceras cordatum J. Sowerby (467)	*Upper Jurassic* Upper Zones of Upper, Oxfordian, Pits of the London Brick Co., Peterborough, England	Calcite & Aragonite (ray 3.36 Å), Aragonite (ray 3.36 Å), Calcite (ray 3.03 Å) & Iron sulfide	❖❖	Pink-creamy

The conchiolin matrices in nacreous layers 123

Species	Age/Location	Mineralogy	Symbols	Description
Prionodoceras sp. (1007)	*Upper Jurassic* Upper Oxfordian, Wiltshire, England	Calcite	❖ sph ves abundant	Dark-brown Crystalline
SUPERFAMILY 21: PERISPHINCTACEAE				
Strenoceras oolithicum (Quenstedt) (409)	*Middle Jurassic* Dogger, Bajocian (Subfurcaten-Schicht), Osterfeld bei Goslar, Harz, Germany	Aragonite & ray 3.03 Å of Calcite Calcite & ray 3.36 Å of Aragonite	cyl varic	Pink-white Chalky
Perisphinctes sp. (1190)	*Jurassic* Kimeridgian, Pas-de-Calais, France	Calcite & Aragonite (1 ray)	❖	Traces of a brown iridescent layer
Perisphinctes (*Virgatosphinctes*) sp. (493)	*Upper Jurassic* U.S.S.R.	Aragonite	am p(?) np(?) mb	Strongly iridescent Nacreous and chalky
Proplanulites koenigii (d'Orbigny) (406)	*Middle Jurassic* Choroschowo, U.S.S.R.	Calcite & Aragonite (ray 3.36 Å)	❖	Iridescent Nacreous & Chalky
Spathia martinsi d'Orb. (1200)	*Upper Jurassic* Callovian, Thouars, France	Calcite	np	White crystalline Rust-coloured coating
Rasenia sp. (537) Sedgwick Museum J 29069	*Jurassic* Kimeridge Clay, Market Rasen, Lincolnshire, England	Aragonite	❖ mb	Iridescent with metallic hues Dark-brown and white Nacreous

Virgatites virgatus (Buch) (511)	*Upper Jurassic* Kimeridgian, Mniownik, Moscow, U.S.S.R.	Carbonate Apatite (Dahlite) & Quartz	❖❖	Iridescent
Virgatites virgatus (Buch) (798)	*Middle Jurassic* Choroschowo, U.S.S.R.	Carbonate Apatite (Dahlite)	❖❖	Strongly iridescent with green metallic hues Crystalline
Virgatites (Zaraskites) sp. (557)	*Jurassic* Ob. Malmstufe, Lower Wolga Layer, Mniownik, Moscow Basin, U.S.S.R.		Sph	
Craspedites subditus (460)	*Upper Jurassic* Aquilonian, Choroschowo, U.S.S.R.	Calcite, Aragonite (ray 3,36 Å) & Iron Sulfide	❖	Iridescent Chalky
Garniericeras catenulatum Fischer de Waldheim (402) (403)	*Upper Jurassic* Malm, Wolga-Stufe, Kaschpur, U.S.S.R.	Aragonite & Calcite (ray 3,03 Å) Calcite & Aragonite Calcite & Aragonite	❖ fl nod sph n p(?)	Strongly iridescent with intense metallic hues Nacreous
(429)	Choroschowo, U.S.S.R.	Carbonate Apatite (Dahlite) Quartz	❖ rib (dahlite, see the text)	Pink-creamy material overlying strongly iridescent, bronze-coloured inner layer. Pink, chalky, powdery, with iridescent green hues

The conchiolin matrices in nacreous layers 125

			cyl varic sl np (septal pattern)	
Leopoldia sp. (1192, 1198)	*Lower Cretaceous* Hauterivian, Gréoux, Alpes de Provence, France	Calcite		Strongly iridescent, creamy & chalky
(484)	Folkestone, England	Aragonite & Calcite (1 ray)		
Platylenticeras (Oxynoticeras) (?) (483)	*Lower Cretaceous* Folkestone, England	Calcite & Aragonite (1 ray)		Strongly iridescent Creamy
SUPERFAMILY 23: HOPLITACEAE				
Acanthohoplites sp. (hannoverensis Jak.?) (495, 556)	*Lower Cretaceous* Lower Albian, Vöhrum and Peine, Hannover, Germany	Aragonite	*	Very brittle Strongly iridescent with intense metallic hues Nacreous & Chalky
Acanthohoplites milletianus d'Orb. (499)	*Lower Cretaceous* Gault, Albian, Folkestone, England	Aragonite	G. 1972b	Pink-white iridescent with violet hues
Acanthohoplites nolani (?) (492)	*Cretaceous* Lower Albian, Senonian-Neocomian, Vöhrum, Hannover, Germany	Aragonite		Strongly iridescent, Pink
Douvilleiceras sp. (844) (septal material)	*Cretaceous* Greensand-Gault Junction Bed, Kent, England		np (septal) angul	Dark-brown Crystalline

Species	Age/Location	Mineralogy	Structure	Appearance
Placenticeras sp. (386: U.S. Geological Survey # 17809; 1001: U.S.G.S. 25483)	*Cretaceous* Coffee Sand, Ratliff, Lee County, Mississippi, USA	Aragonite	✻ and ✻✻ cyl varic mb n p (386) VFG 1971; G 1972b	Very brittle Strongly iridescent with metallic hues Nacreous
Placenticeras sp. (*placenta* De Kay?) (391, 496)	*Upper Cretaceous* Senonian, Bad Lands of Dakota, USA	Aragonite Aragonite & Calcite (ray 3.03 Å) (496)	✻✻ VFG 1971; G 1972b	Iridescent (silvery) Nacreous and chalky-creamy (496)
Placenticeras sp. (U.S. Geological Survey: 27065) (1006)	*Cretaceous* Eutaw Formation, Menabites, Alabama, USA	Aragonite	✻✻ cyl varic mb abundant G 1972b	Strongly iridescent Nacreous
Leymeriella sp. (?) (517)	*Cretaceous* Gault, Albian, Folkestone, England	Aragonite	✻✻✻ mb VFG 1971	Iridescent Slate-coloured & pink-violet Chalky
Hoplites (Anahoplites) splendens J. Sowerby (425, 462, 463)	*Lower Cretaceous* Albian, St. Pô (Wissant), Pas-de-Calais, France	Aragonite		Pink faintly iridescent Chalky
Hoplites tuberculatus J. Sowerby (430) (459)	*Lower Cretaceous* Albian, St. Pô (Wissant), Pas-de-Calais, France	Iron Sulfide & ray 3.36 Å of Aragonite		Pink iridescent with metallic hues Chalky

The conchiolin matrices in nacreous layers

Hoplites auritus Sow. (456)	*Lower Cretaceous* Albian, St. Pô (Wissant), Pas-de-Calais, France	Aragonite & Iron Sulfide	✤ angul	Very brittle Pink iridescent Powdery and Chalky
(479)	*Lower Cretaceous* Albian, Hauterivian, Folkestone, England	Iron Sulfide	✤ fl bro nod	Very brittle Pink iridescent Powdery and chalky
Hoplites jamesoni (625)	*Lower Cretaceous* Algernissen, Germany	Aragonite	ves sph	Iridescent Chalky
Protohoplites ar-chiacianus? (d'Orbigny) (498)	*Lower Cretaceous* Gault, Folkestone, England	Carbonate apatite (Dahlite) Quartz & Iron Sulfide		Iridescent chalky
24 spms including several Hoplitaceae (648-9; 654-659)	*Cretaceous* Middle and Upper Albian, Wealden, Gault, Small-dole, Sussex, England	Aragonite	✤ cyl varic VFG 1971	Brittle Strongly iridescent, pink-violet Nacreous & Chalky
Hoplitaceae sp. (911)	*Cretaceous* Gault, Golden Cap, Seaton, Dorset, England	Aragonite & unidentified ray	✤ VFG 1971	Pink Chalky

SUPERFAMILY 24: ACANTHOCERATACEAE

Oxytropidoceras sp. (Univ. Nacional Ingeniera, Mus. Nacional Paleontol, Lima, Peru, 66215) (1120)	*Lower Cretaceous* Albian, Region alpanira Jauja, Prov. Junin, Peru	Calcite	sph	Black Crystalline

Inflaticeras ventanillensis (1121)	*Lower Cretaceous* Albian, Yauli, Prov. Junín, Peru	Calcite	nod sph sl	Black Crystalline
Collignoniceras sp. (433) Kansas Geol. Survey 10540	*Cretaceous* Gulfian, Blue Hills Shale Member, Carlile Shale, Smith County, Kansas USA	Aragonite	❖❖ fl. n.p. mb very substantial G 1972b	Strongly iridescent with violet metallic hues Nacreous
Texanites sp. (399) U.S. Geol. Survey 27065	*Cretaceous* Upper Eutaw Formation, Marvyn, Russell County, Alabama, USA	Aragonite	❖❖ angul mb	Iridescent Nacreous
Sphenodiscus sp. (389); U.S. Geol. Survey (707–20843)	*Cretaceous* Maestrichtian? Bluffs of Owl Creek Formation, Ripley, Tippah County, Mississippi, USA	Aragonite	❖ Textfig. 7 cyl varic angul sph ves GL 1972b	Brown iridescent Nacreous
Mortoniceras sp.? (918)	*Cretaceous* Gault, Uppermost layer of the cliffs, Charmouth, Dorset, England	Aragonite & Calcite (ray 3.03 Å)	❖ varic	Pink chalky

ABHANDLUNGEN DER AKADEMIE DER WISSENSCHAFTEN UND DER LITERATUR

MATHEMATISCH-NATURWISSENSCHAFTLICHE KLASSE

Jahrgang 1966

1. *Wolfram Ostertag*, Chemische Mutagenese an menschlichen Zellen in Kultur. 124 S., 34 Abb. und 32 Tab. DM 12,–
2. *Ferdinand Claussen* und *Franz Steiner*, Zwillingsforschung zum Rheuma-Problem. 198 S. mit 8 Tab., DM 18,60
3. *Otto H. Schindewolf*, Studien zur Stammesgeschichte der Ammoniten. Lieferung V. 131 S. mit 95 Abb., DM 12,40
4. *Wilhelm Troll* und *Focko Weberling*, Die Infloreszenzen der Caprifoliaceen und ihre systematische Bedeutung. 151 S., 76 Abb., DM 14,20
5. *Hildegard Schiemann*, Über Chondrodystrophie (Achondroplasie, Chondrodysplasie). 61 S. mit 13 Tab. und 19 Abb., DM 5,80
6. *Harm Glashoff*, Endogene Dynamik der Erde und die Diracsche Hypothese. 34 S. mit 9 Abb., DM 4,80
7. *Hubert Forestier* und *Marc Daire*, Anomalies de réactivité chimique aux points de transformation magnétique des corps solides. 15 S. mit 8 Abb., DM 4,80
8. *Otto H. Schindewolf*, Studien zur Stammesgeschichte der Ammoniten. Lieferung VI. 89 S. mit 43 Abb., DM 8,40

Jahrgang 1967

1. *Carl Wurster*, Chemie heute und morgen. 16 S., DM 4,80
2. *Walter Scholz*, Serologische Untersuchungen bei Zwillingen. 26 S. mit 6 Tab., DM 4,80
3. *Pascual Jordan*, Über die Wolkenhülle der Venus. 7 S., DM 4,80
4. *Widukind Lenz*, Lassen sich Mutationen verhüten? 15 S. mit 6 Abb. und 2 Taf., DM 4,80
5. *Otto Haupt* und *Hermann Künneth*, Über Ketten von Systemen von Ordnungscharakteristiken. 24 S., DM 4,80
6. *Klaus Dobat*, Ein bisher unveröffentlichtes botanisches Manuskript Alexander von Humboldts: Über „Ausdünstungs Gefäße" (= Spaltöffnungen) und „Pflanzenanatomie" sowie „Plantae subterraneae Europ. 1794. cum Iconibus". 25 S. mit 13 Abb. und 4 Taf., DM 4,80
7. *Pascual Jordan* und *S. Matsushita*, Zur Theorie der Lie-Tripel-Algebren. 13 S., DM 4,80
8. *Otto H. Schindewolf*, Analyse eines Ammoniten-Gehäuses. 54 S., mit 2 Abb. und 16 Taf., DM 13,–
9. *Adolf Seilacher*, Sedimentationsprozesse in Ammonitengehäusen. 16 S. mit 5 Abb. und 1 Tafel, DM 4,80

Jahrgang 1968

1. *Heinrich Karl Erben*, *G. Flajs* und *A. Siehl*, Über die Schalenstruktur von Monoplacophoren. 24 S. mit 3 Abb. u. 17 Taf., DM 9,–
2. *Pascual Jordan*, Zur Theorie nicht-assoziativer Algebren. 14 S., DM 4,80
3. *Otto H. Schindewolf*, Studien zur Stammesgeschichte der Ammoniten. 181 S. mit 39 Abb. im Text DM 28,40
4. *Heinrich Ristedt*, Zur Revision der Orthoceratidae. 77 S. mit 5 Tafeln, DM 14,–
5. *Pascual Jordan*, *S. Matsushita*, *H. Rühaak*, Über nicht-assoziative Algebren. 19 S. mit 4 Abb., DM 4,80

Jahrgang 1969

1. *Pascual Jordan* und *H. Rühaak*, Neue Beiträge zur Theorie der Lie-Tripel-Algebren und der Osborn-Algebren. 13 S., DM 4,80
2. *Otto Haupt*, Über das Verhalten ebener Bogen in signierten symmetrischen Scheiteln. 32 S., DM 5,–
3. *Pascual Jordan* und *H. Rühaak*, Über einen Zusammenhang der Lie-Tripel-Algebren mit den Osborn-Algebren. 8 S., DM 4,80
4. *Otto H. Schindewolf*, Über den „Typus" in morphologischer und phylogenetischer Biologie. 77 S. mit 10 Abb., DM 12,–
5. *Peter Ax* und *Renate Ax*, Eine Chorda intestinalis bei Turbellarien (*Nematoplana nigrocapitulaAx*) als Modell für die Evolution der Chorda dorsalis. 26 S., DM 4,80
6. *Winfried Haas* und *Hans Mensink*, Asteropyginae aus Afghanistan (Trilobita). 62 S. mit 5 Tafeln und 14 Abb., DM 11,20

Jahrgang 1970

1. *Gerhard Lang*, Die Vegetation der Brindabella Range bei Canberra. Eine pflanzensoziologische Studie aus dem südostaustralischen Hartlaubgebiet. 98 S. mit 18 Abb., 17 Tab. und 10 Fig. auf Taf., DM 20,60
2. *Otto H. Schindewolf*, Stratigraphie und Stratotypus. 134 S. mit 4 Abb., DM 26,–
3. *Hanno Beck*, Germania in Pacifico. Der deutsche Anteil an der Erschließung des Pazifischen Beckens. 95 S. mit 2 Abb. im Text, DM 16,–
4. *Helmut Hutten*, Untersuchung nichtstationärer Austauschvorgänge in gekoppelten Konvektions-Diffusions-Systemen (Ein Beitrag zur theoretischen Behandlung physiologischer Transportprozesse). 58 S. mit 11 Abb., DM 16,–
5. *Anton Castenholz*, Untersuchungen zur funktionellen Morphologie der Endstrombahn. Technik der vitalmikroskopischen Beobachtung und Ergebnisse experimenteller Studien am Iriskreislauf der Albinoratte. 181 S. mit 96 Abb., DM 68,–

Jahrgang 1971

1. *Pascual Jordan*, Diskussionsbemerkungen zur exobiologischen Hypothese. 28 S., DM 4,80
2. *H. K. Erben* und *G. Krampitz*, Eischalen DDT-verseuchter Vögel: Ultrastruktur und organische Substanz. 24 S. mit 12 Tafeln, DM 8,40
3. *Otto H. Schindewolf*, Über Clymenien und andere Cephalopoden. 89 S. mit 10 Abb. und 2 Taf., DM 20,-

Jahrgang 1972

1. *R. Laffitte, W. B. Haberland, H. K. Erben, W. H. Blow, W. Haas, N. F. Hughes, W. H. C. Ramsbottom, P. Rat, H. Tintant, W. Ziegler*, Internationale Übereinkunft über die Grundlagen der Sratigraphie. 24 S., DM 6,20
2. *Karl Hans Wedepohl*, Geochemische Bilanzen. 18 S., DM 4,80
3. *Walter Heitler*, Wahrheit und Richtigkeit in den exakten Wissenschaften. 22 S., DM 4,80
4. *O. Haupt* und *H. Künneth*, Ordnungstreue Erweiterung ebener Bogen und Kurven vom schwachen Ordnungswert Drei. 37 S., DM 10,40
5. *Carl Troll* und *Cornel Braun*, Madrid. Die Wasserversorgung der Stadt durch Qanate im Laufe der Geschichte. 88 S. mit 18 Abb. und 1 Karte, DM 22,-
6. *Heinrich Karl Erben*, Ultrastrukturen und Dicke der Wand pathologischer Eischalen, 26 S. mit 7 Tafeln, DM 12,-
7. *Ernst Hanhart*, Nachprüfung des Erfolges von 30 eugenischen Beratungen bei geplanten Vetternehen. 32 S. m. 3 Abb., DM 8,50
8. *Wolfgang Barnikol*, Zur mathematischen Formulierung und Interpretation von Ventilationsvorgängen in der Lunge. Ein neues Konzept für die Analyse der Ventilationsfunktion. 55 S. mit 9 Abb., DM 16,-

Jahrgang 1973

1. *F. Lotze*, Geologische Karte des Pyrenäisch-Kantabrischen Grenzgebietes. 1:200 000. 22 S. und 3 Faltktn., DM 12,50
2. *Eugen Seibold*, Vom Rand der Kontinente, 23 S. mit 16 Abb., DM 8,60

Jahrgang 1974

1. *O. E. H. Rydbeck*, Radioastronomischer Nachweis von interstellaren CH-Radikalen. 24 S. mit 9 Abb., DM 9,20

Jahrgang 1975

1. *Klaus Bandel*, Embryonalgehäuse karibischer Meso- und Neogastropoden (Mollusca). 175 S. mit 16 Abb., 28 Schemata, 21 Tafeln, DM 48,20
2. *Ingrid Henning*, Die La Sal Mountains, Utah. Ein Beitrag zur Geoökologie der Colorado-Plateau-Provinz und zur vergleichenden Hochgebirgsgeographie. 88 S. mit 14 Abb., 16 Taf. m. 28 Photos, DM 30,60
3. *Wilhelm Lauer*, Vom Wesen der Tropen, Klimaökologische Studien zum Inhalt und zur Abgrenzung eines irdischen Landschaftsgürtels. 52 S. mit 26 Abb., 19 Taf. m. 30 Photos, 1 Faltkte., DM 30,20
4. *Gernot Gräff*, Prüfung der Gültigkeit eines physikalischen Gesetzes. 14 S. mit 3 Abb., DM 6,20

Jahrgang 1976

1. *Walter Heitler*, Über die Komplementarität von lebloser und lebender Materie. 21 S. DM 6,80
2. *Focko Weberling*, Die Pseudostipeln der Sapindaceae. 27 S. mit 11 Abb., DM 9,80
3. *O. Hachenberg* und *U. Mebold*, Die Struktur und der physikalische Zustand des interstellaren Gases aus Beobachtungen der 21 cm HI-Linie. 36 S. mit 16 Abb., DM 18,20

Jahrgang 1977

1. *Dieter Klaus*, Klimafluktuationen in Mexiko seit dem Beginn der meteorologischen Beobachtungsperiode. Studien über Klimaschwankungen und Vegetationsdynamik in Mexiko. Teil I. 81 Seiten mit 25 Abb., DM 28,-
2. *Gert L. Haberland*, Proteolytische Systeme, Steuerungseinheiten im Organismus. 20 Seiten mit 9 Abb., DM 8,-

Jahrgang 1978

1. *Gernot Gräff*, Der Starkeffekt zweiatomiger Moleküle. 18 S. m. 1 Abb., DM 10,-

Jahrgang 1979

1. *Wilhelm Lauer* und *Peter Frankenberg*, Zur Klima- und Vegetationsgeschichte der westlichen Sahara. 61 S. mit 25 Abb., DM 24,40
2. *Günther Ludwig*, Wie kann man durch Physik etwas von der Wirklichkeit erkennen? 16 S., DM 5,20
3. *Günter Lautz*, Miniaturisierung ohne Ende? Entwicklungstendenzen der physikalischen Elektronik. 42 S. mit 38 Abb., DM 14,80
4. *Georg Dhom*, Aufgaben und Ziele einer Krebsforschung am Menschen. 23 S. mit 10 Abb., DM 8,60

Jahrgang 1980

1. *Heinrich Karl Erben*, A Holo-Evolutionistic Conception of Fossil and Contemporaneous Man. 18 S., DM 7,80
2. *Charles Grégoire*, The Conchiolin Matrices in Nacreous Layers of Ammonoids and Fossil Nautiloids: A Survey. Part 1: Shell wall and septa. 128 S. mit 59 Abb., DM 39,20